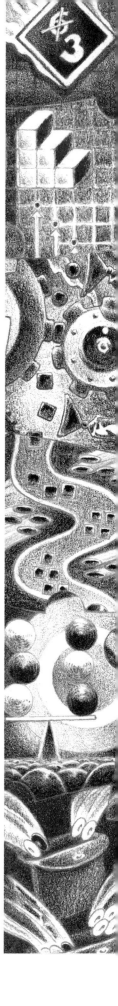

Groundworks

Algebra Puzzles and Problems
Grade 6

Carole Greenes
Carol Findell

D1205208

Creative Publications

Design Director: Gregg McGreevy

Editor: Ann Roper

Art Director: Meg Saint-Loubert

Illustrator: Duane Bibby

Production Coordinator: Cindy Koller

Production Director: Ed Lazar

Designer: Carolyn Deacy Design

Production: University Graphics, Inc.

Manufacturing Coordinator: Michelle Berardinelli

Cover Illustrator: Susan Aiello

Groundworks and Algebra Puzzles and Problems are trademarks of
Creative Publications.

Creative Publications is a registered trademark of Creative Publications.

©1998 Creative Publications® Inc.
Two Prudential Plaza, Suite 1175
Chicago, IL 60601

Printed in the United States of America

ISBN: 0-7622-0558-X

3 4 5 6 7 8 9 10 ML 05 04 03 02 01 00 99

Contents

Teacher Notes

Big Idea: Representation

Big Idea: Proportional Reasoning

Big Idea: Balance

Big Idea: Variable

Big Idea: Function

Big Idea: Inductive Reasoning

Algebra: Puzzles and Problems

Why?

Algebra for Everyone is promoted by the National Council of Teachers of Mathematics (NCTM) in its Curriculum and Evaluation Standards (1989), by the College Board's Equity Project, and by authors of the SCANS Report (1991) on needs of the workplace. As a consequence, school districts around the country require all students to study algebra in high school. In some schools students take formal algebra courses as early as the seventh, or even sixth, grade. Although students are capable of succeeding in algebra, they often do not. Why aren't they successful?

From the authors' experience, even more able math students, jumping from an arithmetic-driven program directly into the study of algebra, often find the new content confusing and daunting. The main reason for this difficulty and for subsequent failure is lack of preparation. Although the NCTM has recommended that students gain experience with the big ideas of algebra during their elementary school years, current mathematics programs do not include such a preparatory program.

What?

Algebra: Puzzles and Problems contains problems for students in grades 4-7, to develop their understanding of six big ideas of algebra:

- ◆ representation
- ◆ proportional reasoning
- ◆ balance
- ◆ variable
- ◆ function
- ◆ inductive/deductive reasoning

The problems capitalize on students' experiences with arithmetic and arithmetic reasoning and help them make the connection between arithmetic and algebra. With such opportunities to make connections and to explore the big ideas of algebra before their formal study of the subject, students enhance their chances for success with algebra and with algebraic reasoning.

Big Ideas

Each problem in the book, though appearing under one big idea, generally requires the understanding of more than one algebraic idea.

Representation

Representation is the display of mathematical relationships in diagrams, drawings, graphs (bar, picto-, line, and scatter plot), symbols, tables, time lines, and text.

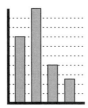 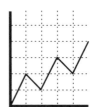 

In this book, *The Point Is* and *Which Graph?* are the families of problems that develop representation ability. Representation ability involves:

◆ interpretation of mathematical relationships shown in many forms

◆ matching of different representations of the same relationship; for example, descriptions in words matched with graphic displays

◆ creation of multiple representations of the same relationship

◆ recognition of how a change in one representation of a relationship effects a change in other representations of the same relationship

Proportional Reasoning

Determining how objects vary in relation to one another is the essence of proportional reasoning. Students use proportional reasoning when they interpret maps and scale drawings, determine dimensions of enlargements and reductions, compute unit costs, generate equivalent ratios, and identify percentages or parts of groups. To give students practice in this type of reasoning, this book includes two families of problems, *Smart Shopping* and *In the Jar.*

Balance

Balance deals with the concept of equality among variable expressions. The problems in *In the Pan* and *Hanging Numbers* aid in the development of understanding of equality and ways to modify inequalities to achieve balance or equality. This is necessary preparation for work with solving equations and inequalities in algebra.

Variable

Fundamental to algebra is understanding the meaning of variable. Using variables is a way of generalizing relationships and representing quantities in formulas, functions, and equations, and of computing in expressions and equations. *What Is Its Weight?, Frames, Logic Grid,* and *Fruit Cocktail* help students to identify relationships among variables, to use variables in systems of equations, and to solve equations with variables, using the processes of substitution and replacement.

Function

Patterns and functions are important ideas in mathematics. A function is a relationship in which two sets are linked by a rule that pairs each element of the first set with exactly one element of the second set. There are three families of problems that deal with functions: *Function Table, Flow Along,* and *Creative Operations.* Students will also learn to describe in words or symbols rules for relating inputs and outputs and to construct inverse operation rules (to undo what other operations did).

Inductive Reasoning

Reasoning inductively requires the ability to examine particular cases, identify patterns and relationships among those cases, and extend the patterns and relationships. In the problem families *Lattice Patterns* and *Shape Teaser,* students learn to recognize, extend, and generalize patterns, and to write rules in words or symbols to describe them.

Deductive Reasoning

Deductive reasoning involves the ability to infer a new relationship from information gleaned from two or more stated or displayed relationships. Most problems in this book require making inferences and drawing conclusions. Deductive reasoning is not one of the big ideas of this book; however, it is often one of the goals for a problem set.

What is in this Book?

Each Algebra: Puzzles and Problems book contains general teacher information, 30 blackline-master problem sets for students (three problems per set), questions, solutions, a Management Chart, and a Certificate of Excellence, both of which are also blackline masters.

Problem Sets

Each problem set consists of four pages. The first page presents the problem. On the facing page, there is teacher information, including goals listing specific mathematical reasoning processes or skills, questions to ask students, and solutions. The third page of each set presents two more problems dealing with the same big idea. Solutions are on the fourth page. Generally there is not room to show more than one solution method for each of these problems. If one method appears on a teacher page, another method may be shown for the second problem. Students can use either method for both problems. There are six problems in each family of problems. The mathematics required for all the problems is in line with the generally approved math curriculum for each grade level.

How to Use this Book

Have students work either individually or in pairs. Because many of the problem types will be new to your students, you may want to have the entire class or a large group of students work on the first problem in a set at the same time. You can use the questions that accompany the problem as the basis for a class discussion. As the students work on the problem, help them with difficulties they may encounter. To help them in their thinking, provide feedback such as, *How do you know? Does that seem reasonable? Explain your answer.* Once students have completed the first problem in each set, you can assign the next two problems for the students to do on their own.

Although the big ideas and the families of problems within them come in a certain order, your students need not complete them in this order. They might work the problems based on the mathematics content of the problems and their alignment with your curriculum, or according to student interests or needs.

One way to use this book is to have students work one set of problems per week for 30 weeks, in what is generally a 36-week school year. If you present and discuss the first problem in a set early in a week, the students could have the rest of the week to complete the other two problems, either in class or as homework.

Since the idea behind the problems is to give students experience with the problems, which focus to a great extent on reasoning and not on computation, it seems reasonable that your students use calculators if they wish to do so.

There is a Management Chart on page vii that you may duplicate for each student to keep in a portfolio. You may award The Certificate of Excellence (page 120) upon the successful completion of problem sets.

Management Chart

Name _____ Class _____ Teacher _____

BIG IDEA	PROBLEM SET				DATE
Representation	The Point Is	A	B	C	
		D	E	F	
	Which Graph?	A	B	C	
		D	E	F	
Proportional Reasoning	Smart Shopping	A	B	C	
		D	E	F	
	In the Jar	A	B	C	
		D	E	F	
Balance	In the Pan	A	B	C	
		D	E	F	
	Hanging Numbers	A	B	C	
		D	E	F	
Variable	What Is Its Weight?	A	B	C	
		D	E	F	
	Frames	A	B	C	
		D	E	F	
	Logic Grid	A	B	C	
		D	E	F	
	Fruit Cocktail	A	B	C	
		D	E	F	
Function	Function Table	A	B	C	
		D	E	F	
	Flow Along	A	B	C	
		D	E	F	
	Creative Operations	A	B	C	
		D	E	F	
Inductive Reasoning	Lattice Patterns	A	B	C	
		D	E	F	
	Shape Teaser	A	B	C	
		D	E	F	

The Point Is (A)

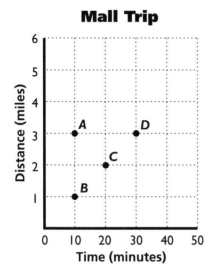

Mall Trip

The time-distance scatter plot represents each person's trip to the mall.

Clues

a. The brothers Jack and Rick live at 43 Broad St.

b. Jack got a ride to the mall; Rick jogged.

c. LaTanya lives closer to the mall than Rick. She biked there.

d. Sarah lives farther from the mall than LaTanya.

Tell which point represents which person.

1 A is _____

2 B is _____

3 C is _____

4 D is _____

The Point Is (A)

Goals
- ◆ Interpret a scatter plot.
- ◆ Match mathematical relationships presented in words with those shown in a graph.
- ◆ Make inferences.

Questions to Ask
- ◆ *What does the vertical axis show?* (Distance in miles)
- ◆ *What do the points on the graph represent?* (Each point represents the distance and the time it took for one person to get to the mall.)
- ◆ *How many minutes did it take C to get to the mall?* (20 minutes)
- ◆ *Who traveled three miles to the mall?* (Both A and D)

Solutions

Clue a Jack and Rick live the same distance from the mall. Therefore, Points A and D at 3 miles represent their trips.

Clue b Since Jack got a ride to the mall, he probably got there faster than Rick, who jogged. Therefore Point A at 10 minutes must represent Jack's trip, and Point D at 30 minutes must represent Rick's trip.

Clue c This clue gives no helpful information.

Clue d Sarah lives farther from the mall than LaTanya, so Point C at 2 miles represents Sarah's trip; Point B at 1 mile represents LaTanya's trip.

1 A is Jack.

2 B is LaTanya.

3 C is Sarah.

4 D is Rick.

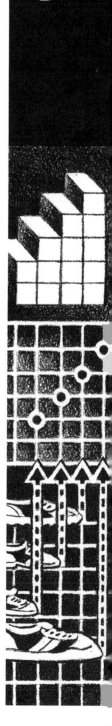

The Point Is (B)

The time-distance scatter plot represents each person's trip to the park.

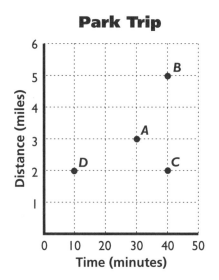

Park Trip

Clues

a. Pedro lives farthest from the park.

b. Tara and Diane live the same distance from the park.

c. Tara got a ride to the park; Diane walked.

d. Ang biked to the park, averaging one mile in ten minutes.

Tell which point represents which person.

1 A is _____   **2** B is _____

3 C is _____   **4** D is _____

The Point Is (C)

The time-distance scatter plot represents each person's trip to the concert.

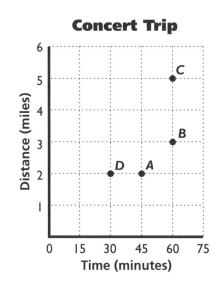

Concert Trip

Clues

a. Marina's trip took half an hour longer than DeWayne's trip.

b. Judy and DeWayne went the same number of miles.

c. Terry averaged 1 mile in 12 minutes.

Tell which point represents which person.

1 A is _____   **2** B is _____

3 C is _____   **4** D is _____

The Point Is (B)

Solutions

Clues a and d identify one certain person. Begin with them, in any order.

Clue a Since Pedro lives farthest from the park, B at 5 miles must represent his trip.

Clue d Ang biked 3 miles in 30 minutes, which is 1 mile in 10 minutes. Point A represents Ang's trip.

Clue b We know that, by elimination, C and D represent Tara's and Diana's trips. This clue offers no new information, but it provides a check. Since the girls live the same distance from the park, C and D, at 2 miles, must represent their trips.

Clue c Since Tara rode to the park, her trip probably took less time than Diane's walk. So D at 10 minutes must represent Tara's trip; C, at 40 minutes, is Diane's trip.

1 A is Ang.

2 B is Pedro.

3 C is Diane.

4 D is Tara.

The Point Is (C)

Solutions

Clue c Terry averaged 1 mile in 12 minutes. Point C shows a trip of 5 miles in 60 minutes, or an average of 1 mile in 12 minutes, and must be his trip.

Clue a Marina's trip took 30 minutes longer than DeWayne's. The only 2 points remaining on the graph that represent trips differing by 30 minutes are B and D. Thus B represents Marina's trip, and D represents DeWayne's.

Clue b This clue offers no helpful information, but it names the fourth person, Judy. Use the clue as a check. Point A represents Judy's trip.

1 A is Judy.

2 B is Marina.

3 C is Terry.

4 D is DeWayne.

The Point Is (D)

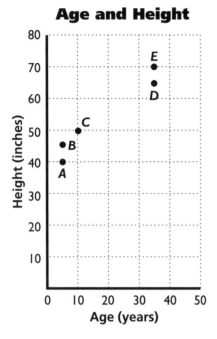

Age and Height

The age-height scatter plot represents each person's age and height. Study it and answer the questions below.

Clues

a. Eric is taller than Emma.

b. Charles is older than Paul.

c. Paul is taller than Arianne.

d. Eric is Paul's father.

e. Emma is Charles's mother.

1 Which point represents Eric's age and height? _____

2 How tall is Charles's mother? _____

3 How many inches taller is Charles than Arianne? _____

4 What are Paul's and Arianne's ages and heights?

The Point Is (D)

Goals
- ◆ Interpret a scatter plot.
- ◆ Match mathematical relationships presented in words with those shown in a graph.
- ◆ Make inferences.

Questions to Ask
- ◆ *What does Point C represent?* (A 10-year-old person, 50 inches tall)
- ◆ *What do Points A and B say about the ages of 2 people?* (They're the same age, about 5.)
- ◆ *Which point represents the tallest person?* (Point E)
- ◆ *How old is the person represented by Point D?* (35 years)
- ◆ *How do you know?* (Point D is halfway between the 30-year line and the 40-year line.)
- ◆ *How much taller is the person represented by Point C than the person represented by Point B?* (5 inches)

Solutions
There is no single clue that identifies one person.

Clues d and e Since Eric is a father and Emma is a mother, they must be the oldest. Their ages and heights must be represented by Points D and E.

Clue a Since Eric is taller than Emma, E represents Eric's age and height and D represents Emma's.

Clue b Charles is older than Paul. Thus Point C must represent Charles, since A and B are the same age.

Clue c Paul is taller than Arianne so Point B represents him and Point A, Arianne.

1 Point E represents Eric's age and height.

2 Charles's mother is Emma who is 65 inches tall.

3 Charles is 10 inches taller than Arianne.

4 Paul is 5 years old and 45 inches tall. Arianne is 5 years old and 40 inches tall.

The Point Is (E)

The age-height scatter plot represents the ages and heights of five people. Study it and answer the questions below.

Clues

a. Rachel is twice as old as Ben.

b. Ben is five inches taller than Hans.

c. Jolene is two feet one inch taller than Ming-Hui.

1 Who is the tallest person? _____

2 How old is Hans? _____

3 Whose age and height are the same numbers? _____

4 Who does Point A represent? _____

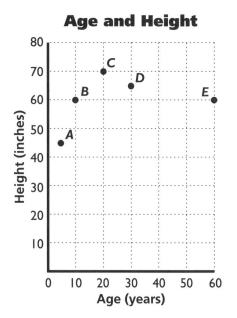

Age and Height

The Point Is (F)

The age-height scatter plot shows the ages and heights of five people. Study it and answer the questions.

Clues

a. Aneka is the youngest person.

b. Thad is older than Maria.

c. Kent is the same height as Aneka.

d. Chen is the same height as Maria.

1 Who does Point D represent? _____

2 How old is Aneka? _____

3 Who is the same age as Kent? _____

4 Who is the tallest person? _____

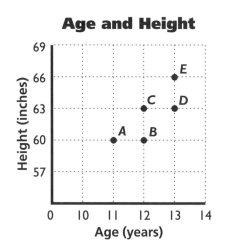

Age and Height

Algebra: Puzzles and Problems

The Point Is (E)

Solutions

Clue c Since Jolene is 2 feet 1 inch, or 25 inches, taller than Ming-Hui, she must be represented by Point C, and Ming-Hui by Point A.

Clue a Since Rachel is twice as old as Ben, Rachel must be Point E, and Ben, age 30, must be Point D. That leaves Point B to represent Hans.

1 Jolene is the tallest person, 70 inches.

2 Hans is 10 years old.

3 Rachel is 60 years old and 60 inches tall.

4 Point A is Ming-Hui.

The Point Is (F)

Solutions

Clue a Aneka, 11 years old and 60 inches tall, is the youngest person. Point A represents her.

Clue c Kent is the same height as Aneka. Point B represents Kent.

Clue d Chen is the same height as Maria, so they must be Points C and D; Thad must be Point E.

Clue b Since Thad, age 13, is older than Maria, she is Point C. If Maria is Point C, then Chen is Point D.

1 Point D represents Chen.

2 Aneka is 11 years old.

3 Maria is the same age as Kent.

4 Thad is the tallest person, 66 inches.

Which Graph? (A)

Graph A

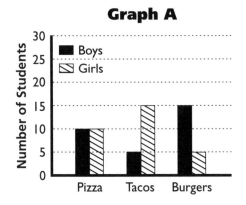

Graph B

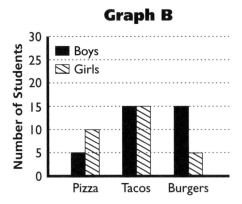

Students chose their favorite fast food: pizza, tacos, or burgers. The double-bar graphs show their choices. Pick the graph, A or B, that shows the relationship.

1 The same number of students chose each food.

Graph _____

2 The number of boys who chose burgers is ten more than the number of boys who chose pizza.

Graph _____

3 The number of boys who chose burgers is equal to the number of boys who did not choose burgers.

Graph _____

4 The same number of boys as girls participated in the survey. Graph _____

Algebra: Puzzles and Problems

Name

Which Graph? (A)

Goals
- ◆ Interpret and compare bar graphs.
- ◆ Match mathematical relationships presented in words with those shown in graphs.

Questions to Ask
- ◆ *In Graph A how many girls chose tacos?* (15)
- ◆ *In Graph A which food was chosen by the same number of boys as girls?* (Pizza)
- ◆ *How can you tell?* (In the double bar labeled Pizza, both bars are the same height.)
- ◆ *How many students chose tacos in Graph B?* (15 boys and 15 girls, 30 in all)
- ◆ *Which food was chosen by 20 students in both graphs?* (Burgers)
- ◆ *In Graph B how many more girls than boys chose pizza?* (5 more)

Solutions

1 Graph A; make a table showing the number of students that chose each food.

Food	Graph A	Graph B
Pizza	10 + 10 = 20	5 + 10 = 15
Tacos	5 + 15 = 20	15 + 15 = 30
Burgers	15 + 5 = 20	15 + 5 = 20

2 Graph B; 15 chose burgers, 5 chose pizza.

3 Graph A; make a table showing the number of boys that chose each food.

Food	Graph A	Graph B
Burgers	15	15
Not burgers	10 + 5 = 15	5 + 15 = 20

4 Graph A; 30 boys and 30 girls.

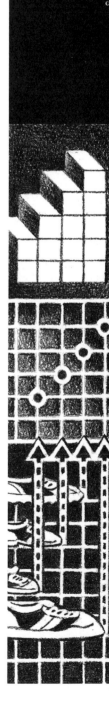

Which Graph? (B)

Students chose their favorite snack food: popcorn, pretzels, peanuts, or chips. The double-bar graphs show the results of the survey.

Pick the graph, A or B, that shows the relationship.

1 There were 90 students in the survey.

Graph _____

2 More boys than girls chose peanuts. Graph _____

3 Thirty-five students chose popcorn. Graph _____

4 One fourth of the students chose peanuts.

Graph _____

Name

Graph A

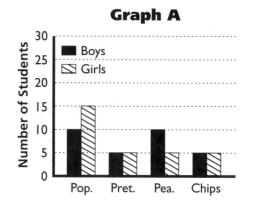

Graph B

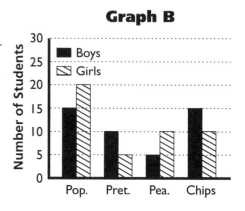

Which Graph? (C)

Students chose their favorite dessert: cake, cookies, ice cream, or fruit. The double-bar graphs show the results of the survey.

Pick the graph, A or B, that shows the relationship.

1 No girls chose cake. Graph _____

2 There were five more girls than boys in the survey.

Graph _____

3 Three times as many girls chose ice cream as chose cookies. Graph _____

4 Half the boys surveyed chose cookies.

Graph _____

Name

Graph A

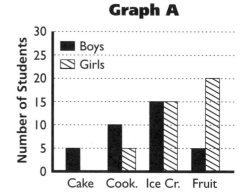

Graph B

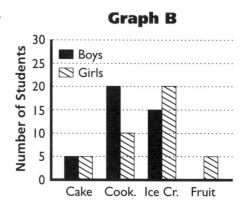

Which Graph? (B)

Solutions

1 Graph B: Make a table to find the total number of students for each graph.

Food	Graph A	Graph B
Popcorn	10 + 15 = 25	15 + 20 = 35
Pretzels	5 + 5 = 10	10 + 5 = 15
Peanuts	10 + 5 = 15	5 + 10 = 15
Chips	5 + 5 = 10	15 + 10 = 25
Total	60	90

2 Graph A

3 Graph B; check the table.

4 Graph A; from the table you know that Graph A represents 60 students, and Graph B represents 90. You know that in Graph A, 15 students chose peanuts; 15 is one fourth of 60.

Which Graph? (C)

Solutions

1 Graph A

2 Graph A; make a table; 40 − 35 = 5.

	Graph A	Graph B
Boys	5 + 10 + 15 + 5 = 35	5 + 20 + 15 = 40
Girls	0 + 5 + 15 + 20 = 40	5 + 10 + 20 + 5 = 40

3 Graph A; five girls chose cookies; 15 girls chose ice cream.

4 Graph B; 20 boys chose cookies, and 5 + 15 + 0, or 20, boys did not.

Algebra: Puzzles and Problems, Grade 6 **11**

Which Graph? (D)

Graph A

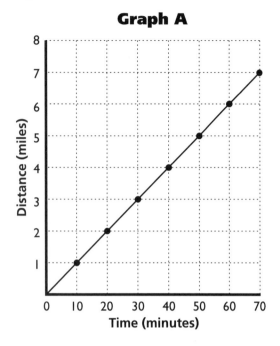

Graph B

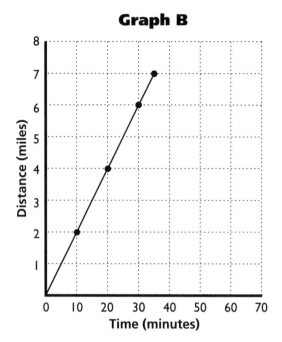

Derek and Scott are brothers. They left home at the same time and rode their bikes to the park. One graph represents Derek's bike ride. The other graph represents Scott's bike ride.

Study the graphs and answer these questions.

1 Derek rode faster than Scott. Which graph represents Derek's bike ride? Graph _____

2 How many minutes did it take Derek to bike to the park? _____

3 How many minutes did it take Scott to bike to the park? _____

4 How far is the park from the boys' house? _____

Name

Which Graph? (D)

Goals
- ◆ Interpret and compare single line graphs.
- ◆ Match mathematical relationships presented in words with those shown in graphs.

Questions to Ask
- ◆ *In Graph A how long did it take the biker to ride 4 miles?* (40 minutes)
- ◆ *In Graph A how far did the biker travel in 60 minutes?* (6 miles)
- ◆ *In Graph B how long did it take the biker to ride 2 miles?* (10 minutes)
- ◆ *In Graph B how far did the biker travel in 15 minutes?* (3 miles)

Solutions
1 Graph B represents Derek's bike ride.

   The graph shows that Derek biked 7 miles to the park in 35 minutes. He averaged 1 mile in 5 minutes. Scott took 70 minutes, averaging 1 mile in 10 minutes. Derek's trip was faster. Compare the slopes of the 2 lines. The slope of the line representing Derek's trip is steeper than the slope showing Scott's trip. A steeper line means a faster trip.

2 Derek biked to the park in 35 minutes.

3 Scott biked to the park in 70 minutes.

4 The park is 7 miles from the boys' home.

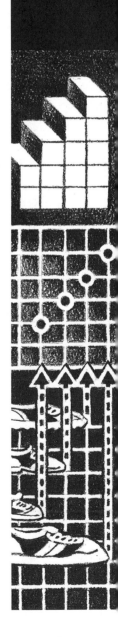

Which Graph? (E)

Ellen and Joan left school and rode their bikes to the library. One graph represents Ellen's bike ride. The other graph represents Joan's ride.

Study the graphs and answer these questions.

1 Ellen averaged 2 miles in 30 minutes. Which graph shows her ride? Graph _____

2 Who rode faster, Ellen or Joan? _____

3 Joan left school at 4:30 P.M. What time did she get to the library? _____

4 How can you tell from the graphs who rode faster?

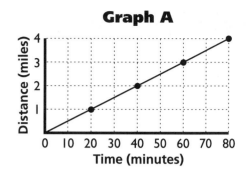

Graph A

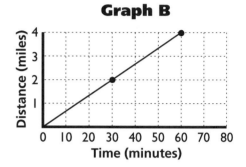

Graph B

Permission is given by the publisher to the purchasing teacher or parent to reproduce this page for classroom or home use only.

Which Graph? (F)

Willie and Jean left the library at the same time and rode their bikes to the swimming pool. One graph represents Willie's bike ride, and the other represents Jean's ride.

Study the graphs and answer these questions.

1 On the way to the pool, Willie stopped for 10 minutes to put air in his bike tires. Jean did not stop. Which graph shows Willie's trip? Graph _____

2 How does the graph show that Willie stopped?

_____

3 Who got to the pool first? _____

4 How far is it from the library to the swimming pool?

_____

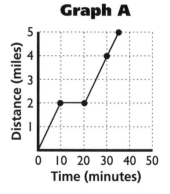

Graph A

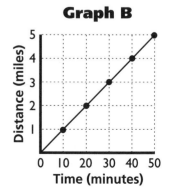

Graph B

Permission is given by the publisher to the purchasing teacher or parent to reproduce this page for classroom or home use only.

Algebra: Puzzles and Problems

Which Graph? (E)

Solutions

1 Graph B represents Ellen's bike ride. The line passes through the points (30, 2) and (60, 4).

2 Ellen biked faster than Joan.

3 Joan arrived at the library at 5:50. Her bike ride took 80 minutes.

80 min = 1 hr 20 min

4 hr 30 min + 1 hr 20 min = 5 hr 50 min

4 The line showing Ellen's ride is steeper than the line representing Joan's ride. Ellen's 4-mile trip took 60 minutes, while Joan's 4-mile trip took 80 minutes.

Which Graph? (F)

Solutions

1 Graph A represents Willie's bike ride.

2 The horizontal part of the graph between 10 and 20 minutes shows the stop. During that time the distance remained 2 miles.

3 Since they left at the same time, and Willie's ride took 35 minutes to Jean's 50, Willie reached the pool first.

4 The distance from the library to the pool is 5 miles.

Algebra: Puzzles and Problems, Grade 6

Smart Shopping (A)

Rulers

Note Pads

3 for $1.20

Office Stuff

Note Pads

4 for $1.50

Two stores sell note pads.

1 Jan said, "I think the note pads at Office Stuff are the better buy." Do you agree? _____ Explain why or why not. How do you think that Jan made her decision?

2 Which store would you go to if you wanted to buy six note pads? _____
Explain your answer.

Name

Smart Shopping (A)

Goals
- ◆ Compare items by identifying the relationship between cost and quantity.
- ◆ Generate equivalent ratios.

Questions to Ask
- ◆ *What is the price of 4 note pads at Office Stuff?* ($1.50)
- ◆ *How many note pads can you get for $1.20 at Rulers?* (3)
- ◆ *Can you tell by just looking at the signs which note pad costs less?* (No) *Why not?* (To compare costs, the numbers of note pads must be the same.)

Solutions

1 Two possible solution methods:

Compute and compare unit costs.
Three pads for $1.20 is $0.40 per pad at Rulers.
Four pads for $1.50 is $1.50 ÷ 4, about $0.38, per pad at Office Stuff. These are less expensive and the better buy.

Equate the prices and compare the numbers of note pads.
The Lowest Common Multiple of $1.20 and $1.50 is $6.
For $6, you get 5 packages, or 15 note pads, at Rulers.
For $6, you get 4 packages, or 16 note pads, at Office Stuff. Since you get more note pads for $6 at Office Stuff, this store has the better buy.

2 Answers will vary. Assume that you can't break the packages up and buy individual pads. Some students may choose 6 note pads for $2.40 at Rulers (even though the unit cost is greater) claiming they want exactly 6 pads.

Other students might rather pay $3 for 8 pads at Office Stuff, saying they want 2 extra pads.

Smart Shopping (B)

Rulers

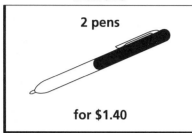

2 pens

for $1.40

Office Stuff

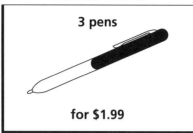

3 pens

for $1.99

Two different stores sell pens.

1 Which store has the better buy, Rulers or Office Stuff? _____ How do you know?

2 Which store would you go to if you want to buy six pens? _____ Why?

Name

Smart Shopping (C)

Rulers

Magnets: 3 packs for $4

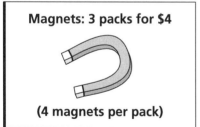

(4 magnets per pack)

Office Stuff

Magnets: 2 packs for $3

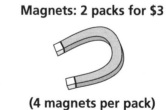

(4 magnets per pack)

These two stores sell magnets.

1 Which store has the better buy on magnets?

_____

2 Tell two different ways to find your answer.

Name

Smart Shopping (B)

Solutions

1 Office Stuff has the better buy.

Compute and compare unit costs.
Rulers: $1.40 ÷ 2 is $0.70 per pen.
Office Stuff: $1.99 ÷ 3 is about $0.66
per pen.

2 Equate the number of pens and
compare prices.
The Lowest Common Multiple of 2 and
3 is 6.
Six pens at Rulers cost 3 × $1.40, or
$4.20.
Six pens at Office Stuff cost 2 × $1.99, or
$3.98, so you would probably go there.

Smart Shopping (C)

Solutions

1 Rulers has the better buy.

2 Three possible solution methods:

Compute and compare unit costs.
At Rulers you get 12 magnets for $4.
That is about $0.33 each. At Office Stuff
you get 2 packages—8 magnets—for $3.
That is about $0.38 each.

Equate the number of magnets and
compare the costs.
The Lowest Common Multiple (LCM)
of 2 and 3 is 6.
Six packages at Rulers cost 2 × $4, or $8.
Six packages at Office Stuff cost 3 × $3,
or $9.

Equate the costs and compare the total
numbers of magnets.
The LCM of $4 and $3 is $12.
At Rulers you get 9 packages, 36
magnets, for $12.
At Office Stuff you get 8 packages, 32
magnets, for $12.
Since 36 is greater than 32, Rulers has
the better buy.

Smart Shopping (D)

Choco's Chips

Cookies

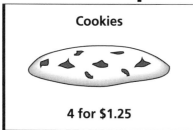

4 for $1.25

Mrs. Fielding's

Cookies

3 for $1.00

Two shops sell chocolate chip cookies.

1 Kelly wants to buy cookies. Which shop has the better buy? _____

2 Explain your answer.

Name

Smart Shopping (D)

Goals
◆ Compare items by identifying the relationship between quantity and cost.

◆ Generate equivalent ratios.

Questions to Ask
◆ *Can you tell which is the better buy by comparing the prices on the signs for each shop?* (No) *Why not?* (Because the number of cookies is not the same)

◆ *Can you tell which is the better buy by comparing the number of cookies sold at each shop?* (No) *Why not?* (Because the prices are not the same)

◆ *How much are 8 cookies at Choco's Chips?* (If 4 cookies cost $1.25, then 8 will cost $2.50.)

Solutions
1 Choco's Chips has the better buy.

2 Three possible solution methods:

Compute the unit cost.
If 4 cookies at Choco's cost $1.25, then 1 costs $1.25 ÷ 4, a little more than 31¢.
If 3 cookies at Mrs. Fielding's cost $1.00, then 1 costs $1.00 ÷ 3, a little more than 33¢.

Equate the numbers of cookies and compare the prices. The LCM of 3 and 4 is 12. Find the prices of 12 cookies.
If 4 cookies at Choco's cost $1.25, then 12 cost 3 × $1.25, or $3.75.
If 3 cookies at Mrs. Fielding's cost $1.00, then 12 cost 4 × $1.00, or $4.

Equate prices and compare the total numbers of cookies.
The LCM of $1.25 and $1.00 is $5.
For $5 you get 16 cookies at Choco's.
For $5 you get 15 cookies at Mrs. Fielding's.

Smart Shopping (E)

Star Store

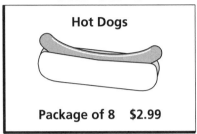

Hot Dogs

Package of 8 $2.99

Stop and Save

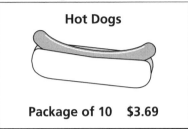

Hot Dogs

Package of 10 $3.69

Two different stores sell hot dogs.

1 Which store has the better buy? _____

2 Explain your answer.

Smart Shopping (F)

Scott's Scoop

Ice Cream

1 pint ice cream $3.49

Izzy's Ices

Ice Cream

$\frac{1}{2}$ gallon ice cream $13.85

Two different shops sell ice cream.

2 pints = 1 quart
2 quarts = $\frac{1}{2}$ gallon

1 Which ice cream shop has the better buy? _____

2 Describe two ways of deciding.

Smart Shopping (E)

Solutions

1 Stop and Save has the better buy.

2 Two possible solution methods:

Make a table for each store and compare prices.

Stop and Save		
Packages	Hot Dogs	Total Cost
1	10	$3.69
2	20	7.38
3	30	11.07
4	40	14.76

Star Store		
Packages	Hot Dogs	Total Cost
1	8	$2.99
2	16	5.98
3	24	8.97
4	32	11.96
5	40	14.95

Forty hot dogs cost less at Stop and Save.

Find the unit prices and compare. One hot dog at Star Store costs $2.99 ÷ 8, or a little more than 37¢. One hot dog at Stop and Save costs $3.69 ÷ 10, or a little less than 37¢.

Smart Shopping (F)

Solutions

1 Izzy's Ices has the better buy.

2 Two possible solution methods:

Compute the cost for a half gallon of ice cream at Scott's Scoop. Compare it with the price at Izzy's Ices.
A half gallon at Scott's Scoop would cost 4 × $3.49, or $13.96. Since $13.96 is more than $13.85, Izzy's Ices has the better buy.

Find the unit prices and compare them. One pint at Izzy's Ices costs $13.85 ÷ 4, a little more than $3.46. This is a little less than the price at Scott's Scoop.

In the Jar (A)

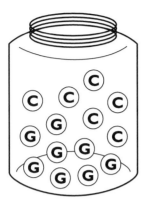

There are some cherry (C) and some grape (G) jelly beans in the jar. Take some beans out of the jar so that for every two cherry beans there is one grape bean in the jar.

1 What is the fewest jelly beans you can take out of the jar? _____

2 How many cherry jelly beans and grape jelly beans are left in the jar? _____

3 Tell how you figured out which jelly beans to leave in the jar.

In the Jar (A)

Goals
- Make drawings or tables to match mathematical relationships presented in words.
- Generate equivalent ratios.

Questions to Ask
- *How many jelly beans are in the jar?* (14)
- *How many jelly beans are cherry?* (Six) *Grape?* (Eight)
- *How many groups of two cherry beans are in the jar?* (Three groups of two)

Solutions
1. Five jelly beans are the fewest.
2. There are six cherry beans and three grape beans left.
3. Students can make drawings or construct tables to solve the problem. Sample drawing:

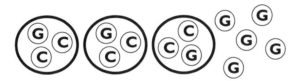

Draw a picture of the jelly beans. Ring groups of three beans: two cherry and one grape. Make three groups and cross out five grape beans.

Sample table:

Number of Cherry Beans	2	4	6
Number of Grape Beans	1	2	3

The table shows the number of cherry jelly beans there must be for one, two, and three grape beans. Since there are only six cherry jelly beans in the jar, you have to take out five grape beans and leave just three.

In the Jar (B)

There are some lemon (L) and some strawberry (S) jelly beans in the jar. Take some jelly beans out so that for every lemon jelly bean there are four strawberry beans left.

1 What is the fewest jelly beans that you have to take out of the jar? _____

2 How many lemon jelly beans and strawberry jelly beans are left in the jar? _____

3 Tell how you got your answers.

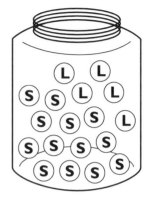

Name

In the Jar (C)

There are some chocolate (C) and some watermelon (W) jelly beans in the jar. Take some jelly beans out so that for every two chocolate jelly beans there are three watermelon beans left.

1 How many different ways could you take beans out of the jar in order to leave two chocolate jelly beans for every three watermelon jelly beans?

_____

2 Explain the different ways you found.

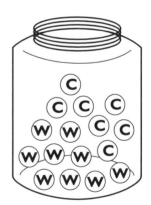

Name

In the Jar (B)

Solutions

1 Take three jelly beans (two lemon and one strawberry) out of the jar.

2 Three lemon and 12 strawberry jelly beans are left.

3 Answers may vary. Students might make a table.

Number of Lemon Beans	1	2	3
Number of Strawberry Beans	4	8	12

Leave three lemon and 12 strawberry beans in the jar. Remove two lemon and one strawberry.

In the Jar (C)

Solutions

1-2 One possible solution method: Make a table.

Left in Jar		Took Out	
Choc.	Wat.	Choc.	Wat.
2	3	5	6
4	6	3	3
6	9	1	0

There are three different ways to remove jelly beans in order to leave the required ratio.

In the Jar (D)

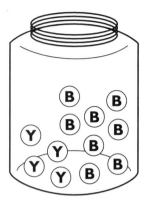

There are some brown (B) and some yellow (Y) Candies in the jar. Take some Candies out so that there are three brown for every two yellow Candies left in the jar.

1 What is the fewest Candies you can take out? _____

2 Draw a picture of the Candies that would be left in the jar. Mark B or Y to show the color of each one.

3 Tell how you decided how many Candies to take out.

Permission is given by the publisher to the purchasing teacher or parent to reproduce this page for classroom or home use only.

Name

In the Jar (D)

Goals
- ◆ Make drawings or tables to match mathematical relationships presented in words.
- ◆ Generate equivalent ratios.

Questions to Ask
- ◆ *Are there more brown or more yellow Candies in the jar?* (More brown)
- ◆ *If there were three brown Candies in the jar, how many yellow Candies would there need to be to have a ratio of three brown to two yellow?* (Two)
- ◆ *Can you leave all eight brown Candies in the jar?* (No)
- ◆ *Why or why not?* (Because the number of brown must be a multiple of three in order to have a ratio of three brown to two yellow)

Solutions
1. The fewest Candies you can take out is two brown.
2. The picture should show six brown and four yellow Candies in the jar.
3. One possible solution method:
   Draw all the Candies that are in the jar.
   Ring a set of three brown and two yellow Candies.
   Ring a second set of three browns and two yellows.
   There are no more yellow Candies, so cross out the two brown Candies that are not in rings.

In the Jar (E)

There are some green (G), some orange (O), and some red (R) gumdrops in the jar. Take some gumdrops out so that for every three green gumdrops, there is one red gumdrop, and one orange gumdrop.

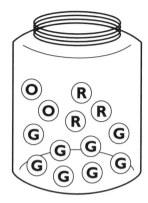

1 What is the fewest gumdrops you can take out?

_____

2 Draw a picture of the gumdrops left in the jar. Mark G, R, or O to show the color of each.

3 Explain how you decided how many gumdrops to take out.

Name

In the Jar (F)

There are some blue (B), some green (G), and some red (R) mints in the jar. Take some mints out so that for every blue mint, there are three red and one green left in the jar.

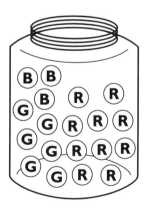

1 What is the fewest mints you can take out? _____

2 How many different ways can you take mints out of the jar so that for every blue mint there are three red and one green left? _____

3 Tell how you figured out your answer to the above question.

Name

Algebra: Puzzles and Problems

In the Jar (E)

Solutions

1 The fewest gumdrops you must remove to leave the required ratio is three: one red and two green.

2 The picture should show six green, two red, and two orange gumdrops left in the jar. One red and two green were removed.

3 One possible solution method is to make a table.

Green	Red	Orange
3	1	1
6	2	2

From the table, students see that they have to stop writing numbers when they get to two orange gumdrops, since that is all there are in the jar.

In the Jar (F)

Solutions

1 The fewest mints you can take out is four: one red and three green.

2 There are three different ways.

3 One possible method is to make a table:

Left in Jar			Took Out		
Blue	Red	Green	Blue	Red	Green
1	3	1	2	7	5
2	6	2	1	4	4
3	9	3	0	1	3

Students will see that there are three different ways to remove mints in order to leave the required ratio.

Algebra: Puzzles and Problems, Grade 6

In the Pan (A)

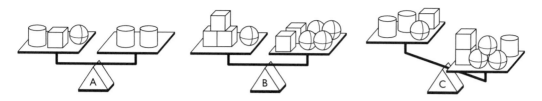

1 Which block, a cylinder, a cube, or a sphere, would you remove from the right pan of Scale C to make it balance? _____

2 Explain or draw how you decided which block to remove.

Algebra: Puzzles and Problems

Name

In the Pan (A)

Goals
- ◆ Deduce the relationship among mass of objects from visual clues.
- ◆ Recognize that balance represents equality.
- ◆ Identify collections of objects having equal mass.
- ◆ Use substitution as a method for equation solving.

Questions to Ask
- ◆ *Which scales show balance?* (A and B)
- ◆ *Which pan of Scale C is heavier?* (The right pan)
- ◆ *Look at Scale A. What would happen if you took one cylinder off the pan on the right?* (The scale would not be balanced. The left pan would be heavier.)
- ◆ *Look at Scale B. What would happen if you took one sphere off the left pan and one sphere off the right pan?* (Scale B would stay balanced because you took the same mass off each side.)

Solutions

1 Remove one sphere from the right pan of Scale C.

2 One possible solution method:

   a: Remove two cubes and one sphere from each pan of Scale B. That leaves one cube on the left pan balancing three spheres on the right. One cube is equal in mass to three spheres.

   b: Remove one cylinder from each side of Scale A. That leaves a cube and a sphere on the left and a cylinder on the right.

   c: Substitute three spheres for the cube on the left of Scale A. Then four spheres on the left balance one cylinder on the right. One cylinder is equal in mass to four spheres.

   d: Substitute four spheres for each cylinder and three spheres for each cube of Scale C. That puts 12 spheres on the left pan and 13 spheres on the right.

   e: Remove one sphere from the right to make Scale C balance.

In the Pan (B)

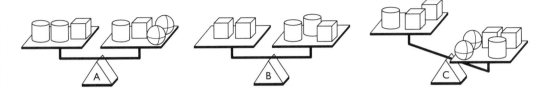

1 Which block, a cylinder, a cube, or a sphere, would you remove from the right pan of Scale C to make it balance? _____

2 Explain or draw how you decided which block to remove.

Name

In the Pan (C)

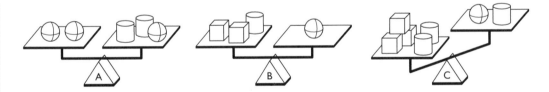

1 Which block, a cylinder, a cube, or a sphere, would you remove from the left pan of Scale C to make it balance? _____

2 Explain or draw how you decided which block to remove.

Name

In The Pan (B)

Solutions

1 Remove one cylinder from the right pan of Scale C.

2 One possible solution method:

a: Remove one cylinder and one cube from each side of Scale A, leaving one cylinder on the left balancing two spheres on the right. One cylinder is equal in mass to two spheres.

b: Remove one cube from each side of Scale B, leaving one cube on the left and two cylinders on the right.

c: Substitute two spheres for each cylinder of Scale B. That leaves one cube on the left balancing four spheres on the right. One cube has mass equal to four spheres.

d: Substitute two spheres for each cylinder and four spheres for each cube on Scale C, leaving 10 spheres on the left and 12 spheres on the right.

e: Make Scale C balance by removing one block from the right pan. The block must have mass equal to two spheres and is a cylinder.

Balance

In The Pan (C)

Solutions

1 Remove one cube from the left pan of Scale C.

2 One possible solution method:

a: One sphere balances a cylinder and two cubes on Scale B.

b: Substitute a cylinder and two cubes for the sphere on the right of Scale C. That makes three cubes and two cylinders on the left pan and two cylinders and two cubes on the right pan.

c: Take one cube from the left pan, leaving the same number of cubes and cylinders on each pan.

Balance

In the Pan (D)

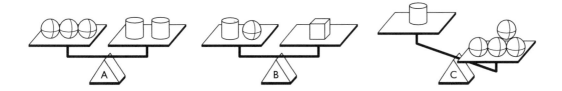

1 Which block, a cylinder, a sphere, or a cube, will balance
Scale C? _____

2 Write or draw the steps you followed to solve the
problem.

Permission is given by the publisher to the purchasing teacher or parent to reproduce this page for classroom or home use only.

Name

In the Pan (D)

Goals
- ◆ Deduce relationships among mass of objects from visual clues.
- ◆ Recognize that balance represents equality.
- ◆ Identify collections of objects having equal mass.
- ◆ Use substitution as a method for equation solving.

Questions to Ask
- ◆ *How many spheres balance two cylinders on Scale A?* (Three)
- ◆ *How many spheres would balance one cylinder?* (One and a half)
- ◆ *How do you know?* (If three spheres balance two cylinders, then, by proportional reasoning, one and a half spheres balance one cylinder.)
- ◆ *On Scale B which block is heavier, the cylinder or the cube?* (The cube)
- ◆ *How do you know?* (Because it alone balances both the cylinder and the sphere)

Solutions
1. Put one cube on the left pan of Scale C to make the scale balance.
2. One possible solution method:

   a: From Scale A you know that two cylinders balance three spheres. Substitute two cylinders for three spheres on the right pan of Scale C, leaving two cylinders and one sphere.

   b: On Scale B one cube balances one cylinder and one sphere. Substitute a cube for one cylinder and one sphere on the right pan of Scale C, leaving one cylinder and one cube.

   c: Put one cube on the left pan of Scale C to balance the scale.

In the Pan (E)

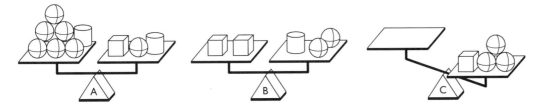

1 Can Scale C be balanced with one block? _____
If so, which block? _____ If not, what is the fewest
blocks that will balance the scale? _____

2 Write or draw the steps you followed to solve the
problem.

Name

In the Pan (F)

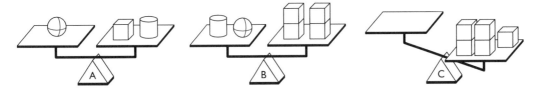

1 What two blocks will balance Scale C? _____

2 Write or draw the steps you followed to identify
the blocks.

Name

Algebra: Puzzles and Problems

In the Pan (E)

Solutions

1 Yes, you can balance Scale C by placing one cylinder on the left pan.

2 One possible solution method:

a: Remove one cylinder and one sphere from each pan of Scale A. That leaves five spheres on the left pan and one cube on the right. One cube balances five spheres.

b: Substitute five spheres for each cube on Scale B. That leaves ten spheres on the left balancing one cylinder and two spheres on the right.

c: Remove two spheres from each pan on Scale B, leaving one cylinder on the right balancing eight spheres on the left. One cylinder is equal in mass to eight spheres.

d: On Scale C the cube is equal in mass to five spheres, so the block on the left has to balance eight spheres. One cylinder is equal in mass to eight spheres, so one cylinder on the left will balance the scale.

Balance

In the Pan (F)

Solutions

1 Put two spheres on the left pan of Scale C to balance the scale.

2 One possible solution method:

a: On Scale B one cylinder and one sphere balance four cubes. Substitute one cylinder and one sphere for four cubes on the right of Scale C. That makes one cube, one cylinder, and one sphere on the right.

b: On Scale A one sphere balances one cylinder and one cube. Put one sphere on the left pan of Scale C to balance the cube and the cylinder on the right.

c: Place another sphere on the left pan of Scale C to balance the sphere on the right.

Balance

Hanging Numbers (A)

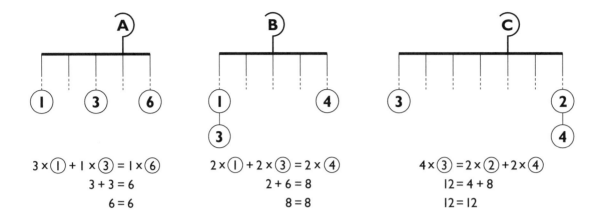

$3 \times \text{①} + 1 \times \text{③} = 1 \times \text{⑥}$
$3 + 3 = 6$
$6 = 6$

$2 \times \text{①} + 2 \times \text{③} = 2 \times \text{④}$
$2 + 6 = 8$
$8 = 8$

$4 \times \text{③} = 2 \times \text{②} + 2 \times \text{④}$
$12 = 4 + 8$
$12 = 12$

The circles represent masses. The number in each circle tells the weight of that mass. The lines on the bars show how far a mass is from the hanger.
These mobiles are balanced.

1 Use all of the numbers given below. Put them in the circles so that this mobile will balance.

2 Tell how you decided where to put the numbers.

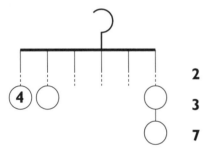

2

3

7

Hanging Numbers (A)

Goals
- ◆ Recognize that equality of moments is represented by a balanced mobile.
- ◆ Compute the moment of each object. (The moment is the product of an object's actual mass and its distance from the hanger or balance point.)

Questions to Ask
- ◆ *In Mobile B how many units to the right of the hanger is the 4-mass?* (2 units)
- ◆ *If the moment of an object is the product of its mass and its distance from the hanger, what is the moment of the 4-mass?* (2×4, or 8)
- ◆ *In Mobile C how many units to the left of the hanger is the 3-mass?* (4 units)
- ◆ *What is its moment?* (4×3, or 12)

Solutions 1

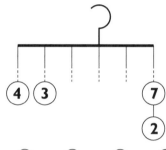

$$3 \times \text{④} + 2 \times \text{③} = 2 \times \text{⑦} + 2 \times \text{②}$$
$$12 + 6 = 14 + 4$$
$$18 = 18$$

2 Use guess-and-check strategy and logical reasoning. Write an equation using the letter M to represent the masses:

$$3 \times 4 + 2 \times M_1 = 2 \times M_2 + 2 \times M_3$$

Since there is a moment of 12 on the left, try the greatest number, 7, on the right. $12 + 2 \times M_1 = 2 \times 7 + 2 \times M_3$. Since the moment on the left (12) is now less than the moment on the right (14), put the greater number, 3, on the left and 2 on the right.

$$12 + 2 \times \text{③} = 14 + 2 \times \text{②}$$

Hanging Numbers (B)

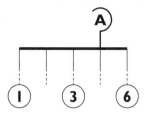

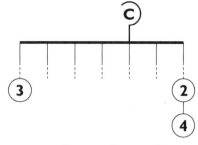

$$3 \times \textcircled{1} + 1 \times \textcircled{3} = 1 \times \textcircled{6}$$
$$3 + 3 = 6$$
$$6 = 6$$

$$2 \times \textcircled{1} + 2 \times \textcircled{3} = 2 \times \textcircled{4}$$
$$2 + 6 = 8$$
$$8 = 8$$

$$4 \times \textcircled{3} = 2 \times \textcircled{2} + 2 \times \textcircled{4}$$
$$12 = 4 + 8$$
$$12 = 12$$

The circles represent masses. The number in each circle tells the weight of that mass. The lines on the bars show how far a mass is from the hanger.

1 Use all of the numbers given. Put them in the circles so that this mobile will balance.

2 Tell how you decided where to put the numbers.

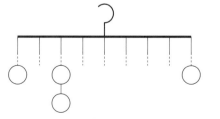

Hanging Numbers (C)

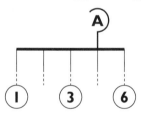

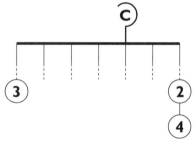

$$3 \times \textcircled{1} + 1 \times \textcircled{3} = 1 \times \textcircled{6}$$
$$3 + 3 = 6$$
$$6 = 6$$

$$2 \times \textcircled{1} + 2 \times \textcircled{3} = 2 \times \textcircled{4}$$
$$2 + 6 = 8$$
$$8 = 8$$

$$4 \times \textcircled{3} = 2 \times \textcircled{2} + 2 \times \textcircled{4}$$
$$12 = 4 + 8$$
$$12 = 12$$

The circles represent masses. The number in each circle tells the weight of that mass. The lines on the bars show how far a mass is from the hanger.

1 Use all of the numbers. Put them in the circles so that this mobile will balance.

2 How did you decide where to put the numbers?

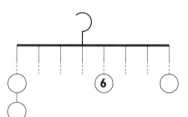

Permission is given by the publisher to the purchasing teacher or parent to reproduce this page for classroom or home use only.

42   **Algebra: Puzzles and Problems**

Hanging Numbers (B)

Solutions

1

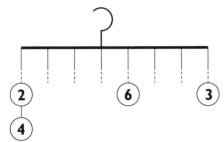

$$3 \times ② + 3 \times ④ = 1 \times ⑥ + 4 \times ③$$
$$6 + 12 = 6 + 12$$
$$18 = 18$$

2 One possible solution method:

There is already a 6-mass 1 unit to the right whose moment is 1×6, or 6.

Write an equation:
$$3 \times M_1 + 3 \times M_2 = 6 + 4 \times M_3$$

To balance 6 on the right, put 2 in the circle on the left to produce a moment of 6.

$$3 \times 2 + 3 \times M_2 = 6 + 4 \times M_3$$

Balance the mobile by placing 4 on the left, 3 units from the hanger, and 3 on the right, 4 units from the hanger.

$$6 + 3 \times ④ = 6 + 4 \times ③$$

Balance

Hanging Numbers (C)

Solutions

1

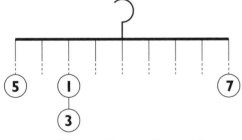

$$4 \times ⑤ + 2 \times ① + 2 \times ③ = 4 \times ⑦$$
$$20 + 2 + 6 = 28$$
$$28 = 28$$

2 One possible solution method:

Write an equation:
$$4 \times M_1 + 2 \times M_2 + 2 \times M_3 = 4 \times M_4$$
Since one mass has to balance all the others, try the greatest number, 7, on the right. $4 \times M_1 + 2 \times M_2 + 2 \times M_3 = 4 \times 7$.

Think of 28 as $20 + 8$ and look for a way to get 20. Try 4×5 and fill in the other numbers.

$$4 \times ⑤ + 2 \times ① + 2 \times ③ = 28$$

Balance

Hanging Numbers (D)

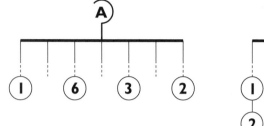

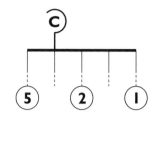

$3 \times \text{①} + 1 \times \text{⑥} = 1 \times \text{③} + 3 \times \text{②}$

$3 + 6 = 3 + 6$

$9 = 9$

$2 \times [\text{①} + \text{②}] = 2 \times \text{③}$

$2 \times 3 = 6$

$6 = 6$

$1 \times \text{⑤} = 1 \times \text{②} + 3 \times \text{①}$

$5 = 2 + 3$

$5 = 5$

The circles represent masses. The number in each circle tells the weight of that mass. The lines on the bars show how far a mass is from the hanger.

These mobiles are balanced.

1 Use all of the numbers given below. Put the numbers in the circles so that this mobile will balance.

2 Describe how you decided where to put the numbers.

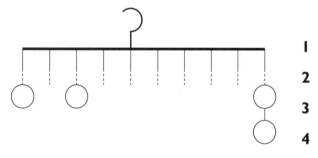

1

2

3

4

Name

Hanging Numbers (D)

Goals
- Recognize that equality of moments is represented by a balanced mobile.
- Compute the moment of each object. (The moment is the product of an object's actual mass and its distance from the hanger or balance point.)

Questions to Ask
- *Look at Mobile A. What do the numbers in the circles represent?* (The weights of the masses)
- *Why does Mobile C balance?* (The distance times the weight of the mass to the left of the hanger [1 × 5] is equal to the distance times the weight of the masses to the right, [2 × 1 + 3 × 1].)
- *In Mobil A what is the product of the distance times the weight of the mass to the left of the hanger?* (3 × 1 + 1 × 6, or 9)
- *To the right of the hanger?* (1 × 3 + 3 × 2, or 9)

Solutions 1

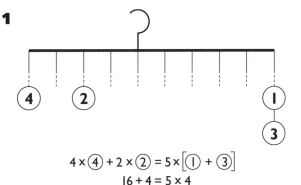

$$4 \times ④ + 2 \times ② = 5 \times \boxed{① + ③}$$
$$16 + 4 = 5 \times 4$$
$$20 = 20$$

2 One possible solution method:

The masses on the right are both 5 units from the hanger. So the moment on the right will be a multiple of 5.

The smallest possible sum for the 2 masses on the right is 1 + 2, or 3. This makes the moment on the right 5 × (① + ②) or 15.

The moment on the left is either 4 × ④ + 2 × ③ = 22 or 4 × ③ + 2 × ④ = 20. Neither matches the moment of 15.

Next try 1 and 3 on the right. The moment is 5 × (① + ③) = 20. Then the moment on the left is 4 × ② + 2 × ④ = 16, or 4 × ④ + 2 × ② = 20, which is the match.

Algebra: Puzzles and Problems, Grade 6 **45**

Hanging Numbers (E)

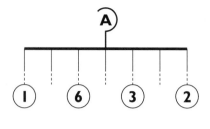

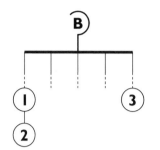

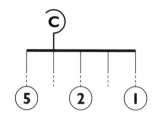

The circles represent masses. The number in each circle tells the weight of that mass. The lines on the bars show how far a mass is from the hanger. These mobiles are balanced.

1 Use all of the numbers given. Put the numbers in the circles so that this mobile will balance.

2 Describe how you decided where to put the numbers.

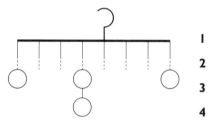

Hanging Numbers (F)

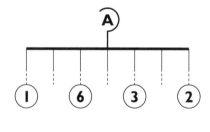

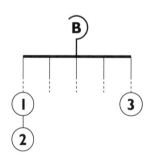

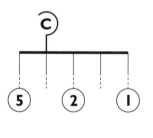

The circles represent masses. The number in each circle tells the weight of that mass. The lines on the bars show how far a mass is from the hanger. These mobiles are balanced.

1 Use all of the numbers given. Put the numbers in the circles so that this mobile will balance.

2 Describe how you decided where to put the numbers.

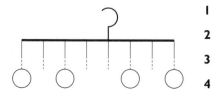

Hanging Numbers (E)

Solutions

1

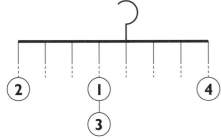

$$4 \times ② + 1 \times \left[① + ③\right] = 3 \times ④$$
$$8 + 1 \times 4 = 12$$
$$8 + 4 = 12$$
$$12 = 12$$

2 One possible solution method:

Try 4, the largest number, on the right side, since there's only 1 number that can go there.

$$4 \times M_1 + 1 \times (M_2 + M_3) = 3 \times 4$$

Then use a guess-and-check strategy to place the rest of the numbers.

Hanging Numbers (F)

Solutions

1

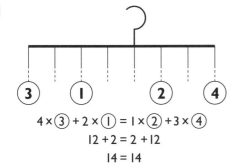

$$4 \times ③ + 2 \times ① = 1 \times ② + 3 \times ④$$
$$12 + 2 = 2 + 12$$
$$14 = 14$$

2 One possible solution method:

Distances from the hanger are the same numbers as the numbers of the masses. Balance 4 on the right with 3 on the left.

$$4 \times 3 + 2 \times M_1 = 1 \times M_2 + 3 \times 4$$

Then use guess-and-check to place the rest of the numbers.

What Is Its Weight? (A)

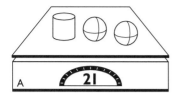 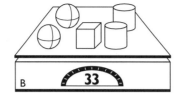 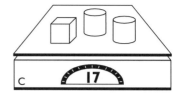

1 Find the weight of each block.

cylinder = _____ pounds

sphere = _____ pounds

cube = _____ pounds

2 Describe how you solved the problem.

What Is Its Weight? (A)

Goals
- ◆ Identify values of blocks from relationships presented symbolically.
- ◆ Replace symbols with numbers to solve problems.
- ◆ Make inferences.

Questions to Ask
- ◆ *What is the total weight of the blocks on Scale A?* (21 pounds)
- ◆ *How is Scale A the same as Scale B?* (All the blocks on Scale A are also on Scale B.)
- ◆ *How could you find the weight of 1 cube and 1 cylinder?* (Subtract 21 pounds, the weight of 1 cylinder and 2 spheres, from Scale B. That leaves 12 pounds, the weight of 1 cube and 1 cylinder.)

Solutions

1 The cylinder weighs 5 pounds.

The sphere weighs 8 pounds.

The cube weighs 7 pounds.

2 One possible solution method:

Look for scales where all the blocks on 1 of them are also on the other.
On Scale A, 1 cylinder and 2 spheres weigh 21 pounds.
If you remove 1 cylinder and 2 spheres from Scale B, the scale will weigh 21 pounds less. Removing these blocks leaves 1 cylinder and 1 cube on Scale B. They weigh 33 – 21, or 12, pounds.
One cylinder and 1 cube on Scale C weigh 12 pounds, so the other cylinder is 17 – 12, or 5, pounds.
If 2 cylinders on Scale C weigh 10 pounds, then the cube must weigh 17 – 10, or 7, pounds.
If the cylinder on Scale A weighs 5 pounds, then the 2 spheres weigh 21 – 5, or 16, pounds. One sphere weighs 16 ÷ 2, or 8, pounds.

What Is Its Weight? (B)

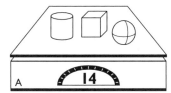 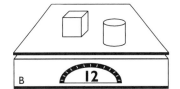 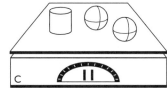

1 Find the weight of each block.

cylinder = _____ pounds

sphere = _____ pounds

cube = _____ pounds

2 Describe how you solved the problem.

What Is Its Weight? (C)

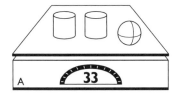 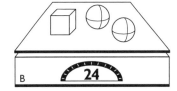 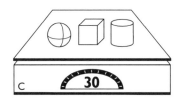

1 Find the weight of each block.

cylinder = _____ pounds

sphere = _____ pounds

cube = _____ pounds

2 Describe how you solved the problem.

Algebra: Puzzles and Problems

What Is Its Weight? (B)

Solutions

1 The cylinder weighs 7 pounds.
The sphere weighs 2 pounds.
The cube weighs 5 pounds.

2 One possible solution method:

On Scale B, 1 cube and 1 cylinder weigh 12 pounds.
If the cube and the cylinder on Scale A weigh 12 pounds, then the sphere weighs 14 – 12, or 2, pounds.
If 2 spheres on Scale C weigh 4 pounds, then the cylinder weighs 11 – 4, or 7, pounds.
If the sphere on Scale A weighs 2 pounds and the cylinder weighs 7, then the cube weighs 14 – 2 – 7, or 5, pounds.

Variable

What Is Its Weight? (C)

Solutions

1 The cylinder weighs 13 pounds.
The sphere weighs 7 pounds.
The cube weighs 10 pounds.

2 One possible solution method:

Both Scales A and C have 1 cylinder and 1 sphere.
However Scale A has a second cylinder and Scale C has a cube. The 3-pound difference in weight means that the cylinder weighs 3 pounds more than the cube.
Use the same reasoning with Scales B and C to show that since both have 1 sphere and 1 cube, the cylinder on Scale C weighs 6 pounds more than the second sphere on Scale B.

Since the cylinder weighs 6 pounds more than the sphere, replacing the sphere on Scale A with a cylinder will add 6 pounds of weight. This will make 3 cylinders weigh 39 pounds; 1 cylinder weighs 13 pounds.
Since the cylinder weighs 3 pounds more than the cube, the cube must weigh 10 pounds. The sphere weighs 6 pounds less than the cylinder, or 7 pounds.

Variable

Algebra: Puzzles and Problems, Grade 6

What Is Its Weight? (D)

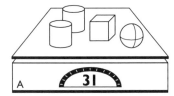 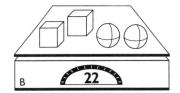 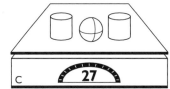

1 Find the weight of each block.

cylinder = _____ pounds

cube = _____ pounds

sphere = _____ pounds

2 Write or draw the steps you followed to find the weight of each block.

Algebra: Puzzles and Problems   © Creative Publications

Name

What Is Its Weight? (D)

Goals
- ◆ Identify values of blocks from relationships presented symbolically.
- ◆ Replace symbols with numbers to solve problems.
- ◆ Make inferences.

Questions to Ask
- ◆ *What blocks are on Scale A?* (2 cylinders, 1 cube, and 1 sphere)
- ◆ *How is Scale A different from Scale C?* (There is a cube on Scale A that is not on Scale C. Also Scale A reads 4 pounds heavier than C.)
- ◆ *If the cylinders on Scale C weigh 12 pounds each, how much does the sphere weigh?* (3 pounds)

Solutions

1 The cylinder weighs 10 pounds.
The cube weighs 4 pounds.
The sphere weighs 7 pounds.

2 One possible solution method:

a: Look for a scale that has all the blocks that are also on another scale. All blocks on Scale C are also on A.

b: The total weight of the blocks on C is 27 pounds.

c: If the C-blocks are removed from A, the cube will weigh 31 − 27, or 4, pounds.

d: If each cube on Scale B is 4 pounds, then the 2 spheres weigh 22 − 8, or 14, pounds. Each sphere is 7 pounds.

e: If the sphere on Scale C is 7 pounds, then the 2 cylinders weigh 20 pounds; each cylinder weighs 20 ÷ 2, or 10, pounds.

What Is Its Weight? (E)

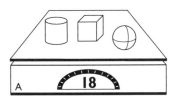 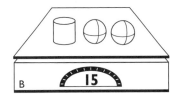 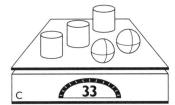

1 Find the weight of each block.

cylinder = _____ pounds

cube = _____ pounds

sphere = _____ pounds

2 Write or draw the steps you followed to find the weight of each block.

Name

What Is Its Weight? (F)

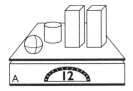 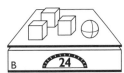 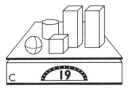 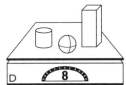

1 Find the weight of each block.

cylinder = _____ pounds

cube = _____ pounds

sphere = _____ pounds

prism = _____ pounds

2 Write or draw the steps you followed to find the weight of each block.

Name

What Is Its Weight? (E)

Solutions

1 The cylinder weighs 9 pounds.
The cube weighs 6 pounds.
The sphere weighs 3 pounds.

2 One possible solution method:

a: All the blocks on Scale B are also on C.

b: The weight of the blocks on Scale B is 15 pounds.

c: If the B-blocks are removed from Scale C, the 2 cylinders left weigh 18 pounds. Each cylinder is 9 pounds.

d: If the cylinder on Scale B is 9 pounds, then the 2 spheres are 15 – 9, or 6, pounds; each sphere is 3 pounds.

e: If the cylinder on Scale A weighs 9 pounds and the sphere 3 pounds, then the cube weighs 18 – 9 – 3, or 6 pounds.

Variable

What Is Its Weight? (F)

Solutions

1 The cylinder weighs 1 pound.
The cube weighs 7 pounds.
The sphere weighs 3 pounds.
The prism weighs 4 pounds.

2 One possible solution method:

If you took the D-blocks off Scale A, the prism would weigh 12 – 8, or 4, pounds. Make a table of values for Scale B.

Cube	1lb	2	3	4	5	6	7
Sphere	21lb	18	15	12	9	6	3

The weight on Scale D is 8 pounds. If the prism is 4 pounds, the table shows the only possible weight for the sphere is 3 pounds. Then the cube weighs 7 pounds. The cylinder weighs 1 pound.

Variable

Algebra: Puzzles and Problems, Grade 6

Frames (A)

$$\square + \square + \triangle = 37$$

$$\square + \triangle = 25$$

Same shapes have same numbers.
Different shapes have different numbers.

1 Write numbers in the shapes to make the equations true.

2 Describe how you found the numbers for Square and for Triangle.

Frames (A)

Goals

◆ Replace variables with numbers in systems of equations.

◆ Identify relationships among variables.

◆ Use substitution as a method for solving systems of equations.

◆ Make inferences.

Questions to Ask

◆ *How many different shapes are in the top equation?* (2)

◆ *If Square equals 15 and Triangle equals 7, will the top equation be true?* (Yes; 15 + 15 + 7 = 37.)

◆ *Would those same values make the bottom equation true?* (No, 15 + 7 = 22, not 25.)

◆ *Would other sets of values make the top equation true?* (Yes)

◆ *What are some other values?* (Square equals 1 and Triangle equals 35; 2, 33; 3, 31; 4, 29; 5, 27; 6, 25; 7, 23; 8, 21; 9, 19; 10, 17; 11, 15; 12, 13; 13, 11; 14, 9; 15, 7; 16, 5; 7, 3; 18, 1)

Solutions

1 Square is 12.

2 Triangle is 13.

2 Two possible solution methods:

Keeping in mind the bottom equation, Square plus Triangle equals 25, replace 1 square and Triangle in the top equation with 25. The remaining square is 37 – 25, or 12.
In the bottom equation, if Square is 12, then Triangle is 25 – 12, or 13.

List pairs of values for Square and Triangle that make the bottom equation true: 0, 25; 1, 24; 2, 23; 4, 21;. . .; 25, 0.
Test these values in the top equation to find those that make that equation true.
The only values for Square and for Triangle that work in both equations are 12 and 13.

Algebra: Puzzles and Problems, Grade 6

Frames (B)

$$\square + \square + \triangle + \triangle = 40$$

$$\square + \square + \triangle = 29$$

Same shapes have same numbers.
Different shapes have different numbers.

1 Write numbers in the shapes to make the equations true.

2 Describe how you found the numbers for Square and for Triangle.

Name

Frames (C)

$$\square + \square + \triangle + \triangle = 30$$

$$\square - \triangle = 5$$

Same shapes have same numbers.
Different shapes have different numbers.

1 Write numbers in the shapes to make the equations true.

2 Describe how you found the numbers for Square and for Triangle.

Name

Frames (B)

Solutions

1 Square is 9.

Triangle is 11.

2 One possible solution method:

Replace the squares and 1 triangle in the top equation with 29, since that is their value in the bottom equation.
The other triangle must be 40 – 29, or 11.
In the bottom equation, if Triangle is 11, then the squares are 18, and Square is 9.

Frames (C)

Solutions

1 Square is 10.

Triangle is 5.

2 One possible solution method:

The bottom equation shows that Square is 5 more than Triangle.
Substitute 2 squares for the 2 triangles in the top equation. Increase the sum by 10, since each square is 5 more than each triangle.
The sum of the new equation is 40 (30 + 5 + 5) and Square is 40 ÷ 4, or 10.
In the bottom equation if Square is 10, Triangle is 10 – 5, or 5.

Frames (D)

$$\square + \triangle + \hexagon = 11$$

$$\hexagon + \hexagon + \triangle = 12$$

$$\hexagon + \triangle = 7$$

Same shapes have same numbers.
Different shapes have different numbers.

1 Write numbers in the shapes to make the equations true.

2 Describe how you found the numbers for Square, for Triangle, and for Hexagon.

Name

Frames (D)

Goals

- ◆ Replace variables with numbers in systems of equations.
- ◆ Identify relationships among variables.
- ◆ Use substitution as a method for solving systems of equations.
- ◆ Make inferences.

Questions to Ask

- ◆ *If Hexagon in the top equation is 3, what is its value in the middle equation?* (It must be 3 also.)
- ◆ *If Square in the top equation is 6, can Hexagon in the bottom equation also be 6?* (No; different shapes must have different numbers.)
- ◆ *In the bottom equation, if Hexagon is 7, what is Triangle?* (It would have to be 0.)

Solutions

1 Square is 4.

Triangle is 2.

Hexagon is 5.

2 One possible solution method:

In the bottom equation Hexagon and Triangle total 7.
Replace Hexagon and Triangle in the top equation with 7.
The new equation reads Square + 7 = 11, so Square is 4.
Replace 1 hexagon and Triangle in the middle equation with 7.
The new equation reads Hexagon + 7 = 12, so Hexagon is 5.
Replace Hexagon with 5 in the bottom equation. If 5 + Triangle = 7, Triangle is 2.

Frames (E)

$$\triangle + \triangle + \square = 17$$

$$\square + \hexagon - \triangle = 8$$

$$\square + \triangle = 11$$

Same shapes have same numbers.
Different shapes have different numbers.

1 Write numbers in the shapes to make the
equations true.

2 Describe how you found the numbers for Square,
for Triangle, and for Hexagon.

Frames (F)

$$\hexagon + \square + \square = 10$$

$$\square + \hexagon = 6$$

$$\triangle + \square + \hexagon = 12$$

Same shapes have same numbers.
Different shapes have different numbers.

1 Write numbers in the shapes to make the
equations true.

2 Describe how you found the numbers for Square,
for Triangle, and for Hexagon.

Algebra: Puzzles and Problems

Frames (E)

Solutions

1 Square is 5.

Triangle is 6.

Hexagon is 9.

2 One possible solution method:

In the bottom equation Square and Triangle total 11.
Replace Square and 1 triangle in the top equation with 11.
The new equation reads Triangle + 11 = 17, so Triangle is 6.
Replace Triangle in the bottom equation with 6.
The new equation reads Square + 6 = 11, so Square is 5.

Replace Triangle with 6 and Square with 5 in the middle equation. Find the value of Hexagon.
If 5 + Hexagon − 6 = 8, Hexagon is 9.

Variable

Frames (F)

Solutions

1 Square is 4.

Triangle is 6.

Hexagon is 2.

2 One possible solution method:

In the middle equation Square and Hexagon total 6.
Replace Hexagon and 1 square in the top equation with 6.
The new equation reads 6 + Square = 10, so Square is 4.
Replace Square with 4 in the middle equation and find the value of Hexagon. Hexagon is 2.
Replace Hexagon with 2 and Square with 4 in the bottom equation. The new equation reads Triangle + 4 + 2 = 12. If Triangle + 6 = 12, Triangle is 6.

Variable

Logic Grid (A)

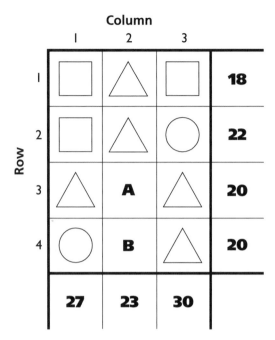

This is a logic grid.

Write numbers in the shapes to make the equations true.

Find the number value of *A* and of *B*.

Same shapes must have same numbers.

Different shapes must have different numbers.

The number at the end of each row and each column is the sum.

1 *A* = _____

2 *B* = _____

3 Describe how you found your answers.

Name

Logic Grid (A)

Goals
- ◆ Replace variables with numbers to make equations true.
- ◆ Identify relationships among variables.
- ◆ Use substitution as a method for solving systems of equations.
- ◆ Make inferences.

Questions to Ask
- ◆ *How many different shapes are in Row 1?* (2)
- ◆ *How many different shapes are in Row 2?* (3)
- ◆ *How many different shapes are in Column 1?* (3)
- ◆ *What is the sum of the shapes in Row 2?* (22)
- ◆ *What is the sum of the shapes and letters in Column 2?* (23)
- ◆ *In Row 3, if A is 10, what would Triangle be?* (Two triangles would be 10, so Triangle would be 5.)

Solutions

1 *A* is 4.

2 *B* is 3.

3 One possible solution method:

In Row 2 we know that the sum of the 3 shapes is 22. These 3 shapes appear in Column 1. You know Square + Triangle + Circle = 22, so replace those 3 shapes in Column 1 with 22. The other square is 27 − 22, or 5.
Replace these same 3 shapes in Column 3 with 22. The other triangle is 30 − 22, or 8.
Replace Square with 5 and Triangle with 8 in Row 2. Circle is 22 − (8 + 5), or 9.
Replace each triangle in Row 3 with 8. *A* is 20 − 8 − 8, or 4.
Replace Circle and Triangle in Row 4 with 9 and 8 respectively. Then *B* is 20 − (9 + 8), or 3.

Logic Grid (B)

This is a logic grid.
Write numbers in the shapes to make the equations true.
Find the number value of A and of B.

Same shapes must have same numbers.
Different shapes must have different numbers.
The number at the end of each row and each column is the sum.

1 A = _____

2 B = _____

3 Describe how you found your answers.

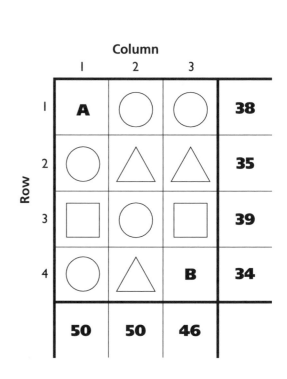

Column

	1	2	3	
1	A	△	☐	24
2	☐	☐	☐	27
3	◯	△	◯	27
4	B	△	◯	22
	32	**30**	**38**	

Row

Logic Grid (C)

This is a logic grid.
Write numbers in the shapes to make the equations true.
Find the number value of A and of B.

Same shapes must have same numbers.
Different shapes must have different numbers.
The number at the end of each row and each column is the sum.

1 A = _____

2 B = _____

3 Describe how you found your answers.

Column

	1	2	3	
1	A	◯	◯	38
2	◯	△	△	35
3	☐	◯	☐	39
4	◯	△	B	34
	50	**50**	**46**	

Row

Logic Grid (B)

Solutions

1 *A* is 8.

2 *B* is 5.

3 One possible solution method:

In Row 2, 3 squares equal 27, so Square is 9.

Replace each square in Column 3 with 9. The 2 circles equal 38 − (9 + 9), or 20. Circle equals 20 ÷ 2, or 10.

Replace Square in Column 2 with 9. Three triangles equal 30 − 9, or 21; Triangle is 7.

Replace Triangle and Square in Row 1 with 7 and 9. Then *A* is 24 − (7 + 9), or 8.

Replace Triangle and Circle in Row 4 with 7 and 10. Then *B* is 22 − (10 + 7), or 5.

Logic Grid (C)

Solutions

1 *A* is 8.

2 *B* is 9.

3 One possible solution method:

In Column 2, 2 circles and 2 triangles total 50. So Circle + Triangle must be 25.

In Row 2 if Circle and 1 triangle are 25, then the other triangle must be 35 − 25, or 10.

In Row 2 if Triangle is 10, then Circle is 35 − 10 − 10, or 15.

In Row 3 if Circle is 15, 2 squares must be 39 − 15, or 24; Square is 24 ÷ 2, or 12.

In Row 1 if Circle is 15, then *A* is 38 − 15 − 15, or 8.

In Row 4 if Circle is 15 and Triangle is 10, then *B* is 34 − 15 − 10, or 9.

Logic Grid (D)

Column

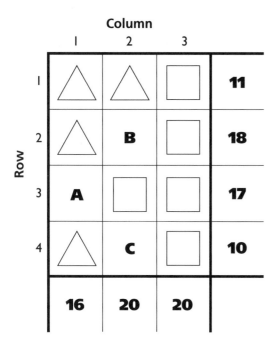

This is a logic grid.

Write numbers in the shapes to make the equations true.

Find the number value of A, of B, and of C.

Same shapes must have same numbers.

Different shapes must have different numbers.

The number at the end of each row and each column is the sum.

1 A = _____

2 B = _____

3 C = _____

4 Describe how you found your answers.

Name

Logic Grid (D)

Goals
- ◆ Replace variables with numbers in systems of equations.
- ◆ Identify relationships among variables.
- ◆ Use substitution as a method for solving systems of equations.
- ◆ Make inferences.

Questions to Ask
- ◆ *What is in Row 3?* (The letter A and 2 squares)
- ◆ *What is in Column 2?* (1 triangle, 1 square, and the letters B and C)
- ◆ *What is the sum of the shapes and the letter in Row 2?* (18)
- ◆ *What is the sum of the shapes and the letter in Column 1?* (16)
- ◆ *To figure out the value of Square, which row or column would you look at first?* (Column 3, which has only squares and you can find the value of each one)

Solutions

1 $A = 7$

2 $B = 10$

3 $C = 2$

4 One possible solution method:

Look for a row or a column in which all of the shapes are the same.
Column 3 has only squares.
Since 4 squares total 20, Square must be 5.
In Row 3 replace each square with 5.
Then $A + 5 + 5 = 17$, and A is 7.
In Column 1 if A is 7, the 3 triangles total 9, and Triangle is 3.
In Row 4 replace Triangle with 3 and Square with 5.
The equation would be $3 + C + 5 = 10$; $C = 2$.
In Row 2 replace Triangle with 3 and Square with 5. Solve for B:
$3 + B + 5 = 18$; $B = 10$.

Logic Grid (E)

This is a logic grid.
Write numbers in the shapes to make the equations true.
Find the number value of A, of B, and of C.

Same shapes must have same numbers.
Different shapes must have different numbers.
The number at the end of each row and each column is the sum.

1 A = _____

2 B = _____

3 C = _____

4 Describe how you found your answers.

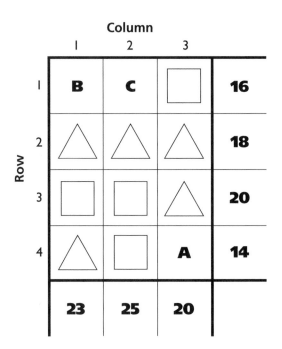

Logic Grid (F)

This is a logic grid.
Write numbers in the shapes to make the equations true.
Find the number value of A, of B, and of C.

Same shapes must have same numbers.
Different shapes must have different numbers.
The number at the end of each row and each column is the sum.

1 A = _____

2 B = _____

3 C = _____

4 Describe how you found your answers.

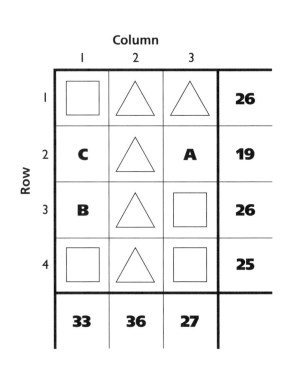

Logic Grid (E)

Solutions

1 $A = 1$

2 $B = 4$

3 $C = 5$

4 One possible solution method:

Look for a row or column which has all the same shapes.
Row 2 has all triangles, so each triangle is $18 \div 3$, or 6.
Replace Triangle in Row 3 with 6. The equation now reads Square + Square + 6 = 20.
The 2 squares must be 14, so Square is 7.
Replace Triangle with 6 and Square with 7 in Row 4. Solve for A: $6 + 7 + A = 14$; $13 + A = 14$; $A = 1$.

Replace Square with 7 and each triangle with 6 in Column 1 and solve for B: $B + 6 + 7 + 6 = 23$; $B = 4$.
Replace B with 4 and Square with 7 in Row 1. Solve for C. $4 + C + 7 = 16$; $11 + C = 16$; $C = 5$.

Variable

Logic Grid (F)

Solutions

1 $A = 2$

2 $B = 9$

3 $C = 8$

4 One possible solution method:

In Row 4 you see that 2 squares and Triangle total 25.
Replace the 2 squares and Triangle in Column 3 with 25.
Then $A + 25 = 27$ and $A = 2$.
Column 2 has 4 triangles, so Triangle is 9.
Replace A with 2 and Triangle with 9 in Row 2.
Then $C + 9 + 2 = 19$; $C + 11 = 19$; $C = 8$.

Replace each triangle in Row 1 with 9; solve for Square. Square is 8.
Replace each square and C with 8 in Column 1. Solve for B: $8 + 8 + B + 8 = 33$; $24 + B = 33$; $B = 9$.

Variable

Algebra: Puzzles and Problems, Grade 6

Fruit Cocktail (A)

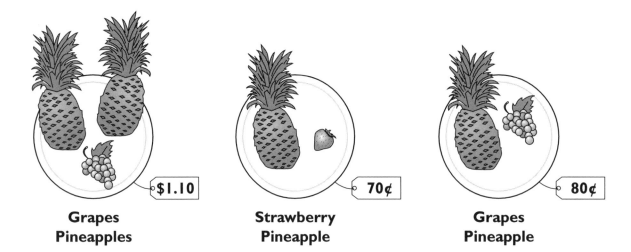

**Grapes   Strawberry   Grapes
Pineapples   Pineapple   Pineapple**

$1.10   70¢   80¢

Find the price of each fruit.
Same fruits have same prices.
Different fruits have different prices.

1 How much is one bunch of grapes? _____

2 How much is a pineapple? _____

3 How much is the strawberry? _____

4 Explain how you found the prices.

Name

Fruit Cocktail (A)

Goals
- ◆ Identify relationships among variables.
- ◆ Replace variables with numbers in systems of equations.
- ◆ Use substitution as a method for solving systems of equations.
- ◆ Make inferences.

Questions to Ask
- ◆ *What fruit is on the plate that costs $1.10?* (Two pineapples and some grapes)
- ◆ *Does another plate have some of the same fruit on it?* (Yes, the third plate has one pineapple and some grapes.)
- ◆ *What fruit costs 70¢?* (A pineapple and a strawberry)
- ◆ *If the strawberry costs 60¢, how much is the pineapple?* (10¢)
- ◆ *How do you know?* (Together they cost 70¢.)

Solutions
1. One bunch of grapes costs 50¢.
2. A pineapple costs 30¢.
3. The strawberry costs 40¢.
4. One possible solution method:
   - **a:** The third plate shows that a bunch of grapes and one pineapple cost 80¢.
   - **b:** Replace one pineapple and the grapes on the first plate with 80¢. The remaining pineapple must cost $1.10 – $0.80, or $0.30.
   - **c:** If the pineapple on the third plate is 30¢, then the grapes must cost 80¢ – 30¢, or 50¢.
   - **d:** If the pineapple on the middle plate is 30¢, the strawberry costs 70¢ – 30¢, or 40¢.

Fruit Cocktail (B)

Grapes
Pineapple
$1.40

Strawberries
Pineapple
70¢

Strawberry
Pineapple
45¢

Find the price of each fruit.
Same fruits have same prices.
Different fruits have different prices.

1 How much is one bunch of grapes? _____

2 How much is a pineapple? _____

3 How much is a strawberry? _____

4 Explain how you found the prices.

Permission is given by the publisher to the purchasing teacher or parent to reproduce this page for classroom or home use only.

Fruit Cocktail (C)

Grapes
Strawberry
Pineapple
$1.00

Strawberry
Pineapple
50¢

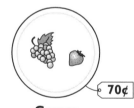

Grapes
Strawberry
70¢

Find the price of each fruit.
Same fruits have same prices.
Different fruits have different prices.

1 How much is one bunch of grapes? _____

2 How much is a pineapple? _____

3 How much is a strawberry? _____

4 Explain how you found the prices.

Algebra: Puzzles and Problems

Permission is given by the publisher to the purchasing teacher or parent to reproduce this page for classroom or home use only.

Fruit Cocktail (B)

Solutions

1 One bunch of grapes costs 60¢.

2 A pineapple costs 20¢.

3 A strawberry costs 25¢.

4 One possible solution method:

a: On the third plate a strawberry and a pineapple cost 45¢.

b: Replace one strawberry and the pineapple on the middle plate with 45¢. The other strawberry must be 70¢ – 45¢, or 25¢.

c: If the strawberry on the third plate is 25¢, then the pineapple is 45¢ – 25¢, or 20¢.

d: If the pineapple on the first plate is 20¢, then the two bunches of grapes are $1.40 – $0.20, or $1.20. One bunch of grapes is $1.20 ÷ 2, or $0.60.

Fruit Cocktail (C)

Solutions

1 One bunch of grapes costs 50¢.

2 A pineapple costs 30¢.

3 A strawberry costs 20¢.

4 One possible solution method:

a: On the middle plate a strawberry and a pineapple cost 50¢.

b: Substitute 50¢ for the strawberry and the pineapple on the first plate. That means that the grapes cost $1.00 – $0.50, or $0.50.

c: On the third plate the grapes and strawberry cost 70¢. If the grapes cost 50¢, then a strawberry costs 20¢.

d: Substitute 70¢ for the grapes and strawberry on the first plate. The pineapple costs $1.00 – $0.70, or $0.30.

Variable

Fruit Cocktail (D)

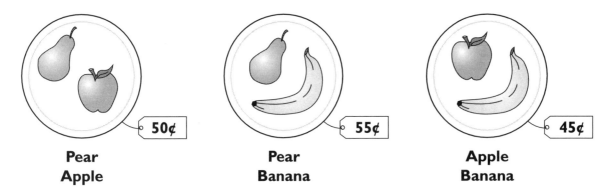

Pear
Apple

50¢

Pear
Banana

55¢

Apple
Banana

45¢

Find the price of each fruit.
Same fruits have same prices.
Different fruits have different prices.

1 How much is a pear? _____

2 How much is an apple? _____

3 How much is a banana? _____

4 Explain how you found the price of a pear.

Fruit Cocktail (D)

Goals
- ◆ Identify relationships among variables.
- ◆ Replace variables with numbers in systems of equations.
- ◆ Use substitution as a method for solving systems of equations.
- ◆ Make inferences.

Questions to Ask
- ◆ *Look at the first plate. What could the price of the apple and of the pear be?* (The apple could cost anywhere from 1¢ to 49¢. The price of the pear would be the difference between 50¢ and the price of the apple.)
- ◆ *If the pear on the first plate is 10¢, how much would the apple cost?* (40¢)
- ◆ *Could the pear cost 10¢?* (No) *Why or why not?* (Look at the middle plate. If the pear is 10¢, the banana would be 45¢. Then the banana on the third plate would be 45¢, and the apple wouldn't cost anything.)

Solutions
1. A pear costs 30¢.
2. An apple costs 20¢.
3. A banana costs 25¢.
4. One possible solution method:
   - **a:** Compare the first and middle plates. Each one has a pear. The differences between the plates are: The first plate has an apple; the middle plate has a banana, and costs 5¢ more than the first plate.
   - **b:** Since the middle plate costs more, we know that the banana costs 5¢ more than the apple.
   - **c:** Substitute a banana for the apple on the third plate; increase the cost by 5¢ to 50¢.
   - **d:** The third plate now has two bananas for 50¢, so one banana is 25¢.
   - **e:** Replace the banana on the middle plate with 25¢ and compute the price of the pear, which is 30¢.
   - **f:** Replace the pear on the first plate with 30¢ and compute the price of the apple, which is 20¢.

Fruit Cocktail (E)

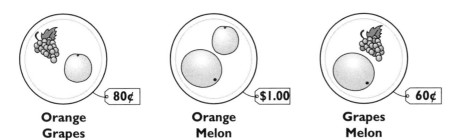

Orange   **Orange**   **Grapes**
Grapes   **Melon**   **Melon**

Find the price of each fruit.
Same fruits have same prices.
Different fruits have different prices.

1 How much is an orange? _____

2 How much is a melon? _____

3 How much is a bunch of grapes? _____

4 Explain how you found the price of the grapes.

Fruit Cocktail (F)

Kiwi   **Peach**   **Kiwi**
Peach   **Lemon**   **Lemon**

Find the price of each fruit.
Same fruits have same prices.
Different fruits have different prices.

1 How much is a kiwi? _____

2 How much is a peach? _____

3 How much is a lemon? _____

4 Explain how you found the price of a lemon.

78   *Algebra: Puzzles and Problems*

Fruit Cocktail (E)

Solutions

1 An orange costs 60¢.

2 A melon costs 40¢.

3 A bunch of grapes costs 20¢.

4 One possible solution method:

a: Compare the first and the middle plates. The oranges are the same price, so the melon must be 20¢ more than the grapes.

b: Substitute a melon for grapes on the third plate and increase the cost by 20¢ to 80¢.

c: Find the cost of 1 melon. 80¢ ÷ 2 is 40¢.

d: Replace the melon on the middle plate with 40¢. Compute the price of the orange, which is $1 – $0.40, or $0.60.

e: Replace the orange with 60¢ on the first plate. The grapes cost 80¢ – 60¢, or 20¢.

Variable

Fruit Cocktail (F)

Solutions

1 A kiwi costs 45¢.

2 A peach costs 35¢.

3 A lemon costs 20¢.

4 One possible solution method:

a: Compare the first and middle plates. There is a peach on each plate. The kiwi must cost 25¢ more than the lemon.

b: Substitute a kiwi for the lemon on the third plate. Increase the cost by 25¢ to 90¢.

c: The third plate now has 2 kiwis for 90¢. One kiwi is 45¢.

d: Replace the kiwi on the first plate with 45¢ and find the price of the peach. The peach is 35¢.

e: Replace the peach on the middle plate with 35¢. Compute the price of the lemon. The lemon is 20¢.

Variable

Function Table (A)

Input	Output
0	4
1	7
2	10
3	13
4	16
6	
	25
8	
9	
	34
	64
40	

The table shows some inputs and some outputs.

1 Complete the table.

2 Use words to write a rule for finding the output when you know the input.

3 Use words to write a rule for finding the input when you know the output.

Algebra: Puzzles and Problems

© Creative Publications

Name

Function Table (A)

Goals
- ◆ Describe in words a rule that relates output to input.
- ◆ Describe in words a rule that relates input to output.
- ◆ Use inverse operations to compute inputs when given outputs.

Questions to Ask
- ◆ *What is the output when the input is 0?* (4)
- ◆ *What is the input when the output is 13?* (3)
- ◆ *What rules can you apply to input 1, to get output 7?* (Add 6; multiply by 2, add 5; multiply by 3, add 4; etc.)
- ◆ *Which rule gives output 10 for input 2?* (Multiply input by 3, add 4.)

Solutions

1

Input	Output
0	4
1	7
2	10
3	13
4	16
6	22
7	25
8	28
9	31
10	34
20	64
40	124

2 Rule: Multiply input by three, add four to the product.

One possible solution method:

The outputs are greater than the inputs so the rule probably involves multiplication, addition, or both.
List rules that would give seven as an output for an input of one.
(See Questions to Ask, above.)
Test the rules with the other inputs and outputs.

3 Subtract four from the output and divide the difference by three to get the input.

Algebra: Puzzles and Problems, Grade 6

Function Table (B)

The table shows some inputs and some outputs.

1 Complete the table.

2 Use words to write a rule for finding the output when you know the input.

3 Use words to write a rule for finding the input when you know the output.

Input	Output
1	0
2	3
3	8
4	15
5	
6	
	48
	63
	99
11	
12	
	399

Permission is given by the publisher to the purchasing teacher or parent to reproduce this page for classroom or home use only.

Function Table (C)

The table shows some inputs and some outputs.

1 Complete the table.

2 Use words to write a rule for finding the output when you know the input.

3 Use words to write a rule for finding the input when you know the output.

Input	Output
0	1
1	1.5
2	2
3	2.5
4	3
5	
6	
7	
	5
	6
	9
20	

Permission is given by the publisher to the purchasing teacher or parent to reproduce this page for classroom or home use only.

Algebra: Puzzles and Problems

Function Table (B)

Solutions

1

Input	Output
1	0
2	3
3	8
4	15
5	24
6	35
7	48
8	63
10	99
11	120
12	143
20	399

2 Rule: Multiply the input by itself and subtract one from the product.

One strategy involves thinking of a rule that relates an input to its corresponding output, then testing rules with other inputs and outputs until finding one that always works.

Each output is one less than a square number. (0 is one less than 1; 3 is one less than 4; 8 is one less than 9, etc.)

3 Add one to the output and find the square root of the sum.

Function

Function Tables (C)

Solutions

1

Input	Output
0	1
1	1.5
2	2
3	2.5
4	3
5	3.5
6	4
7	4.5
8	5
10	6
16	9
20	11

2 One possible solution method and rule: Divide input by two, add one to the quotient.

Since there are decimals, particularly 0.5 or one half, in some outputs, think about dividing by 2 as part of the rule. Try dividing by 2 on some inputs, and compare the quotients with the outputs. For example, $5 \div 2 = 2.5$. You need 3.5, so adding 1 gets you there. Try the rule on six, $6 \div 2 = 3$. You need 4, so add 1. The rule seems to work. Test it some more.

3 Subtract one from the output and multiply by two.

Function

Algebra: Puzzles and Problems, Grade 6

Function Table (D)

Input	Output
0	−1
1	2
2	5
3	8
4	11
	14
9	
10	
15	
	59
	74
	89

The table shows some inputs and some outputs.

1 Complete the table.

2 Use words to write a rule for finding the output when you know the input.

3 Use words to write a rule for finding the input when you know the output.

Algebra: Puzzles and Problems

Permission is given by the publisher to the purchasing teacher or parent to reproduce this page for classroom or home use only.

Name

Function Table (D)

Goals
- ◆ Describe in words a rule that relates outputs to inputs.
- ◆ Describe in words a rule that relates inputs to outputs.
- ◆ Use inverse operations to compute inputs when given outputs.

Questions to Ask
- ◆ *What is the output for input 0?* ($^-$1)
- ◆ *What is the input for output 5?* (2)
- ◆ *What are some ways to get from input 1 to output 2?* (Add 1 to the input; multiply the input by 2; multiply the input by 3, subtract 1; square the input, add 1; etc.)

Solutions

1

Input	Output
0	$^-$1
1	2
2	5
3	8
4	11
5	14
9	26
10	29
15	44
20	59
25	74
30	89

2 One possible solution method and rule:

Multiply the input by three and subtract one.

Look at differences between consecutive outputs. The differences are always three. Try multiplying the inputs by three. Look at the outputs. They are one less. So subtract one.
(Example: Input $2 \times 3 = 6$; $6 - 1 =$ Output 5)

3 One possible rule: Add one to the output and divide by three.

Function Table (E)

The table shows some inputs and some outputs.

1 Complete the table.

2 Use words to write a rule for finding the output when you know the input.

3 Use words to write a rule for finding the input when you know the output.

Input	Output
0	0
1	1
2	8
3	27
4	64
	125
6	
7	
8	
	1,000
	1,331
	8,000

Name

Function Table (F)

The table shows some inputs and some outputs.

1 Complete the table.

2 Use words to write a rule for finding the output when you know the input.

3 Use words to write a rule for finding the input when you know the output.

Input	Output
0	0
1	2
2	6
3	12
4	20
	30
	42
	90
	110
20	
22	
24	

Name

Function Table (E)

Solutions

1

Input	Output
0	0
1	1
2	8
3	27
4	64
5	125
6	216
7	343
8	512
10	1,000
11	1,331
20	8,000

2 One possible solution method and rule:
Input times input times input equals the output.

Since the numbers get large fast, try multiplying:
Any number $\times 0 = 0$; only $1 \times 1 = 1$; $2 \times 4 = 8$; $3 \times 9 = 27$.
Four and 9 are square numbers. How do you get from 2 to 8? Multiply by 2 again.

3 One possible rule: The input is the cube root of the output.

Function Table (F)

Solutions

1

Input	Output
0	0
1	2
2	6
3	12
4	20
5	30
6	42
9	90
10	110
20	420
22	506
24	600

2 One possible rule:
Multiply the input by a number that is one more than the input. Input $\times$ (Input + 1) = Output.

3 One possible rule: Subtract the input from the output and find the square root of the difference. Or, find the closest square number less than the output. Find its square root; that is the input.

Algebra: Puzzles and Problems, Grade 6

Flow Along (A)

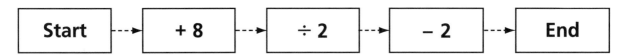

When the Start number is 2, the End number is 3. For each
Start number, give the End number.

For example, when the Start number is 2, here's how you
get to 3, the End number:

(Start) $2 + 8 = 10$; $10 \div 2 = 5$; $5 - 2 = 3$ (End)

1 Start: 0   End: _____

2 Start: 1   End: _____

3 Start: 6   End: _____

When the End number is 3, the Start number is 2.

For each End number, give the Start number.

For example, when the End number is 3, here's how you
get to 2, the Start number:

(End) $3 + 2 = 5$; $5 \times 2 = 10$; $10 - 8 = 2$ (Start)

4 End: 7   Start: _____

5 End: 8   Start: _____

6 End: 12   Start: _____

Name

Flow Along (A)

Goals
- ◆ Use inverse operations.
- ◆ Follow and complete sequences of computations.

Questions to Ask

- ◆ *Suppose that the Start number is 4. Follow the arrows; what do you do to 4 first?* (Add 8 to get 12.)
- ◆ *What happens to 12?* (Divide it by 2 to get 6.)
- ◆ *What happens to 6?* (Subtract 2 to get 4.)
- ◆ *What is the End number?* (4)
- ◆ *Suppose that the End number is 20. How could you get to the Start number?* (Work backwards; use the opposite operations.)
- ◆ *What would you do first?* (Add 2 to get 22)
- ◆ *What number would you have to divide by 2 to get 22?* (44; multiply 22 by 2)
- ◆ *To what number did you have to add 8 to get 44?* (44 − 8, or 36)
- ◆ *What is the Start number?* (36)

Solutions

To get the End number, add 8 to the Start number, divide the sum by 2, and subtract 2 from the quotient.

1 End: 2 (0 + 8 = 8; 8 ÷ 2 = 4; 4 − 2 = 2)

2 End: 2.5 (1 + 8 = 9; 9 ÷ 2 = 4.5; 4.5 − 2 = 2.5)

3 End: 5 (6 + 8 = 14; 14 ÷ 2 = 7; 7 − 2 = 5)

To get the Start number, add 2 to the End number, multiply the sum by 2, and subtract 8 from the product.

4 Start: 10 (7 + 2 = 9; 2 × 9 = 18; 18 − 8 = 10)

5 Start: 12 (8 + 2 = 10; 2 × 10 = 20; 20 − 8 = 12)

6 Start: 20 (12 + 2 = 14; 2 × 14 = 28; 28 − 8 = 20)

Flow Along (B)

| Start | ⟶ | × 2 | ⟶ | ÷ 2 | ⟶ | + 2 | ⟶ | End |

When the Start number is 7, the End number is 9. For each
Start number, give the End number.
For example: (Start) 7 × 2 = 14; 14 ÷ 2 = 7; 7 + 2 = 9 (End).

1 Start: 1   End: _____

2 Start: 5   End: _____

3 Start: 2   End: _____

When the End number is 8, the Start number is 6.
For each End number, give the Start number.

4 End: 6   Start: _____

5 End: 12   Start: _____

6 End: 18   Start: _____

Name

Flow Along (C)

| Start | ⟶ | + 6 | ⟶ | × 5 | ⟶ | − 4 | ⟶ | End |

When the Start number is 5, the End number is 51. For each
Start number, give the End number.

1 Start: 0   End: _____

2 Start: 3   End: _____

3 Start: 10   End: _____

When the End number is 56, the Start number is 6.
For each End number, give the Start number.
For example: (End) 56 + 4 = 60; 60 ÷ 5 = 12; 12 − 6 = 6 (Start).

4 End: 31   Start: _____

5 End: 46   Start: _____

6 End: 21   Start: _____

Name

Algebra: Puzzles and Problems

Flow Along (B)

Solutions

To get the End number, multiply the Start number by 2, divide that product by 2, and add 2 to the quotient.

Some students may discover that multiplying and then dividing by 2 are inverse operations, so to get the End number, all they need to do is add 2 to the Start number.)

1 End: 3
$(1 \times 2 = 2; 2 \div 2 = 1; 1 + 2 = 3)$

2 End:
$7 (5 \times 2 = 10; 10 \div 2 = 5; 5 + 2 = 7)$

3 End:
$4 (2 \times 2 = 4; 4 \div 2 = 2; 2 + 2 = 4)$

To get the Start number, use inverse operations. Subtract 2 from the End number, multiply that difference by 2, and divide that product by 2. (Or just subtract 2 from the End number.)

4 Start: 4
$(6 - 2 = 4; 4 \times 2 = 8; 8 \div 2 = 4)$

5 Start: 10
$(12 - 2 = 10; 10 \times 2 = 20; 20 \div 2 = 10)$

6 Start: 16
$(18 - 2 = 16; 16 \times 2 = 32; 32 \div 2 = 16)$

Function

Flow Along (C)

Solutions

To get the End number, add 6 to the Start number, multiply the sum by 5, and subtract 4 from the product.

1 End: 26
$(0 + 6 = 6; 6 \times 5 = 30; 30 - 4 = 26)$

2 End: 41
$(3 + 6 = 9; 9 \times 5 = 45; 45 - 4 = 41)$

3 End: 76
$(10 + 6 = 16; 16 \times 5 = 80; 80 - 4 = 76)$

To get the Start number, add 4 to the End number, divide the sum by 5, and subtract 6 from the quotient.

4 Start: 1
$(31 + 4 = 35; 35 \div 5 = 7; 7 - 6 = 1)$

5 Start: 4
$(46 + 4 = 50; 50 \div 5 = 10; 10 - 6 = 4)$

6 Start: $^-1$
$(21 + 4 = 25; 25 \div 5 = 5; 5 - 6 = {}^-1)$

Function

Flow Along (D)

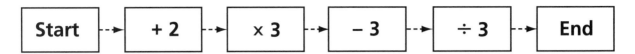

| Start | → | + 2 | → | x 3 | → | – 3 | → | ÷ 3 | → | End |

When the Start number is 1, the End number is 2.
For each Start number, give the End number.

1 Start: 2   End: _____

2 Start: 5   End: _____

3 Start: 10   End: _____

When the End number is 13, the Start number is 12.
For each End number, give the Start number.

4 End: 17   Start: _____

5 End: 25   Start: _____

6 End: 49   Start: _____

Name

Flow Along (D)

Goals
- ◆ Use inverse operations.
- ◆ Follow and complete sequences of computations.

Questions to Ask
- ◆ *If the Start number is 0, what will the End number be?* (1)
- ◆ *If the End number is 4, what will the Start number be?* (3)
- ◆ *How do you get the Start number when you know the End number?* (You work backwards, doing inverse operations.)

Solutions

1 End: 3 ($2 + 2 = 4$; $3 \times 4 = 12$; $12 - 3 = 9$; $9 \div 3 = 3$)

2 End: 6 ($5 + 2 = 7$; $7 \times 3 = 21$; $21 - 3 = 18$; $18 \div 3 = 6$)

3 End: 11 ($10 + 2 = 12$; $3 \times 12 = 36$; $36 - 3 = 33$; $33 \div 3 = 11$)

Some students may notice that each End number is 1 more than the Start number. A simpler flow chart (Add 1 to each input.) would work.

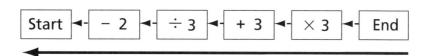

Start ◄- – 2 ◄- ÷ 3 ◄- + 3 ◄- × 3 ◄- End

Students might draw a chart to show the backwards operations needed to find the Start number, given the End number. Above is such a chart.

4 Start: 16 ($17 \times 3 = 51$; $51 + 3 = 54$; $54 \div 3 = 18$; $18 - 2 = 16$)

5 Start: 24 ($25 \times 3 = 75$; $75 + 3 = 78$; $78 \div 3 = 26$; $26 - 2 = 24$)

6 Start: 48 ($49 \times 3 = 147$; $147 + 3 = 150$; $150 \div 3 = 50$; $50 - 2 = 48$)

Again, some students may discover a simpler flow chart. End number minus 1 will give the Start number.

Flow Along (E)

| Start | → | + 4 | → | × 4 | → | – 2 | → | ÷ 2 | → | End |

When the Start number is 1, the End number is 9. For each Start number, give the End number.

1 Start: 2

End: _____

2 Start: 5

End: _____

3 Start: 12

End: _____

When the End number is 13, the Start number is 3. For each End number, give the Start number.

4 End: 19

Start: _____

5 End: 23

Start: _____

6 End: 49

Start: _____

Flow Along (F)

| Start | → | – 2 | → | × 4 | → | + 4 | → | ÷ 2 | → | End |

When the Start number is 1, the End number is 0. For each Start number, give the End number.

1 Start: 2

End: _____

2 Start: 6

End: _____

3 Start: 10

End: _____

When the End number is 12, the Start number is 7. For each End number, give the Start number.

4 End: 20

Start: _____

5 End: 24

Start: _____

6 End: 50

Start: _____

Flow Along (E)

Solutions

1 End: 11
(2 + 4 = 6; 6 × 4 = 24; 24 − 2 = 22;
22 ÷ 2 = 11)

2 End: 17
(5 + 4 = 9; 9 × 4 = 36; 36 − 2 = 34;
34 ÷ 2 = 17)

3 End: 31
(12 + 4 = 16; 16 × 4 = 64; 64 − 2 = 62;
62 ÷ 2 = 31)

A simpler flow chart is × 2 and +7 which
gives the same output for each input.

4 Start: 6
(19 × 2 = 38; 38 + 2 = 40; 40 ÷ 4 = 10;
10 − 4 = 6)

5 Start: 8
(23 × 2 = 46; 46 + 2 = 48; 48 ÷ 4 = 12;
12 − 4 = 8)

6 Start: 21
(49 × 2 = 98; 98 + 2 = 100; 100 ÷ 4 =
25; 25 − 4 = 21)

A simpler flow chart is −7 and ÷2 which
gives the same input for each output.

Function

Flow Along (F)

Solutions

1 End: 2
(2 − 2 = 0; 0 × 4 = 0; 0 + 4 = 4;
4 ÷ 2 = 2)

2 End: 10
(6 − 2 = 4; 4 × 4 = 16; 16 + 4 = 20;
20 ÷ 2 = 10)

3 End: 18
(10 − 2 = 8; 8 × 4 = 32; 32 + 4 = 36;
36 ÷ 2 = 18)

A simpler flow chart is × 2 and −2 which
gives the same output for each input.

4 Start: 11
(20 × 2 = 40; 40 − 4 = 36; 36 ÷ 4 = 9;
9 + 2 = 11)

5 Start: 13
(24 × 2 = 48; 48 − 4 = 44; 44 ÷ 4 = 11;
11 + 2 = 13)

6 Start: 26
(50 × 2 = 100; 100 − 4 = 96; 96 ÷ 4 =
24; 24 + 2 = 26)

A simpler flow chart is +2 and ÷2 which
gives the same input for each output.

Function

Algebra: Puzzles and Problems, Grade 6

Creative Operations (A)

What operations does △ do?

Look at these examples.

A 6 △ 0 = 12

B 4 △ 3 = 11

C 10 △ 1 = 21

D 8 △ 3 = 19

Solve these problems.

1 5 △ 0 = _____

2 9 △ 2 = _____

3 3 △ _____ = 10

4 _____ △ 7 = 11

5 Explain what operations △ stands for.

6 Tell how you solved Problem 3.

Name

Creative Operations (A)

Goals
◆ Identify relationships among sets of three numbers.

◆ Generate a rule that describes the relationship.

Questions to Ask
◆ *What is the answer in Example C?* (21)

◆ *Why is the pyramid there?* (It represents some number operations.)

◆ *What does the pyramid do to 10 and to 1 in Example C?* (It helps them equal 21.)

◆ *How could it do this?* (It could mean double the first number and add the second number; it multiplies 10 by 2 and adds 1.)

◆ *What is the first number in Example B?* (4)

◆ *What is the second number?* (3) *What is the answer?* (11)

◆ *Would the same operations as above work in Example B, too?* (Yes; $4 \times 2 = 8$; $8 + 3 = 11$.)

Solutions
1 $5 \times 2 + 0 = \mathbf{10}$

2 $9 \times 2 + 2 = \mathbf{20}$

3 $3 \times 2 + \mathbf{4} = 10$

4 $\mathbf{2} \times 2 + 7 = 11$

5 Pyramid means: Double the first number and add the second number. ($2 \times$ First Number + Second Number = Answer)

6 Show Problem 3 as: $2 \times 3 + SN = 10$.
Solve for SN (Second Number). $6 + SN = 10$. $SN = 4$.

Creative Operations (B)

What operations does ⬡ do? Look at these examples.

A 2 ⬡ 5 = 11

B 0 ⬡ 4 = 1

C 10 ⬡ 3 = 31

D 6 ⬡ 6 = 37

Solve these problems.

1 4 ⬡ 2 = _____

2 3 ⬡ 5 = _____

3 7 ⬡ _____ = 57

4 _____ ⬡ 9 = 19

5 Explain what operations ⬡ stands for.

6 Tell how you solved Problem 4.

Name

Creative Operations (C)

What operations does △ do? Look at these examples.

A 3 △ 4 = 5

B 8 △ 10 = 13

C 9 △ 2 = 10

D 14 △ 7 = 17.5

Solve these problems.

1 6 △ 8 = _____

2 4 △ 5 = _____

3 _____ △ 9 = 9.5

4 10 △ _____ = 16

5 Explain what operations △ stands for.

6 Tell how you solved Problem 4.

Name

Algebra: Puzzles and Problems

Creative Operations (B)

Solutions

1 $4 \times 2 + 1 = $ **9**

2 $3 \times 5 + 1 = $ **16**

3 $7 \times$ **8** $+ 1 = 57$

4 **2** $\times 9 + 1 = 19$

5 Cube means: Multiply the first number by the second number and add 1. (First Number $\times$ Second Number $+ 1 = $ Answer)

6 Show Problem 4 as: $FN \times 9 + 1 = 19$. Solve for FN (First Number). $FN \times 9 = 18$. $FN = 2$.

Function

Creative Operations (C)

Solutions

1 $6 + (8 \div 2) = $ **10**

2 $4 + (5 \div 2) = $ **6.5**

3 **5** $+ (9 \div 2) = 9.5$

4 $10 + ($ **12** $\div 2) = 16$

5 Cone means: Add the first number to half of the second number. (First Number $+$ (Second Number $\div 2) = $ Answer)

6 Think of Problem 4 in this way: To get 16, you must add 6 to 10; 6 is half of 12, the second number.

Function

Creative Operations (D)

What operations does △ do?

Look at these examples.

A $2 \triangle 4 = 3$

B $3 \triangle 7 = 5$

C $5 \triangle 4 = 4.5$

D $0 \triangle 6 = 3$

Solve these problems.

1 $6 \triangle 10 = $ _____

2 $20 \triangle 25 = $ _____

3 $8 \triangle $ _____ $= 10$

4 _____ $\triangle 9 = 8.5$

5 Explain what operations △ stands for.

6 Tell how you solved Problem 3.

Name

Creative Operations (D)

Goals
- ◆ Identify relationships among sets of three numbers.
- ◆ Generate a rule that describes the relationship.

Questions to Ask
- ◆ *What is the relationship between the numbers and the answer in the first example?* (2 pyramid 4 gives an answer between 2 and 4.)
- ◆ *What can you say about the answer number in each example?* (It is always between the other two numbers.)
- ◆ *How could you use two numbers to get an answer that is between them?* (The mean of two numbers is always between them.)

Solutions
1. Mean of 6 and 10 = **8**
2. Mean of 20 and 25 = **22.5**
3. Mean of 8 and **12** = 10
4. Mean of **8** and 9 = 8.5
5. Pyramid means: Add the first and second numbers and find their mean. This is the answer, the number exactly between the two.
6. Show Problem 3 as: $(8 + SN) \div 2 = 10$. Solve for SN (Second Number). If half of $8 + SN$ is 10, then $8 + SN$ is 20, or twice as much. $8 + SN = 2 \times 10$; $8 + SN = 20$; $SN = 12$.

Creative Operations (E)

What operations does ⬡ do?

Look at these examples.

A 5 ⬡ 4 = 6

B 8 ⬡ 2 = 14

C 3 ⬡ 2 = 4

D 3 ⬡ 6 = 0

E 4 ⬡ 5 = 3

F 8 ⬡ 8 = 8

G 2 ⬡ 3 = 1

H 5 ⬡ 5 = 5

Solve these problems.

1 1 ⬡ 0 = _____

2 5 ⬡ 2 = _____

3 6 ⬡ _____ = 10

4 _____ ⬡ 3 = 5

5 Explain what operations ⬡ stands for.

6 Tell how you solved Problem 4.

Name

Creative Operations (F)

What operations does △ do?

Look at these examples.

A 3 △ 6 = 0

B 8 △ 6 = 5

C 100 △ 100 = 50

D 5 △ 5 = 2.5

E 6 △ 5 = 3.5

F 6 △ 8 = 2

G 100 △ 50 = 75

H 6 △ 1 = 5.5

Solve these problems.

1 6 △ 2 = _____

2 8 △ 4 = _____

3 5 △ _____ = 1

4 _____ △ 4 = 2

5 Explain what operations △ stands for.

6 Tell how you solved Problem 3.

Name

Creative Operations (E)

Solutions

1 $(1 \times 2) - 0 = \mathbf{2}$

2 $(5 \times 2) - 2 = \mathbf{8}$

3 $(6 \times 2) - \mathbf{2} = 10$

4 $(\mathbf{4} \times 2) - 3 = 5$

5 Cube means: Multiply the first number by 2 and subtract the second number from the product. Or, subtract the second number from the first and add the result to the first number.

6 Show Problem 4 as: $(2 \times FN) - 3 = 5$. Solve for FN (First Number). $2 \times FN = 8$. $FN = 4$.

Function

Creative Operations (F)

Solutions

1 $6 - (2 \div 2) = \mathbf{5}$

2 $8 - (4 \div 2) = \mathbf{6}$

3 $5 - (\mathbf{8} \div 2) = 1$

4 $\mathbf{4} - (4 \div 2) = 2$

5 Cone means: Subtract half of the second number from the first. Or, double the first number, subtract the second number from the product, and divide by 2.

6 Show Problem 3 as: $(2 \times 5 - SN) \div 2 = 1$. Solve for SN (Second Number). $(10 - SN) \div 2 = 1$. $10 - SN = 2$. $SN = 8$.

Function

Algebra: Puzzles and Problems, Grade 6

Lattice Patterns (A)

Columns

	A	B	C	D	E	F
Row 1	1	2	3	4	5	6
Row 2	7	8	9	10	11	12
Row 3	13	14	15	16	17	18
Row 4	19	20	21	22	23	24
Row 5	25	26	27	28	29	30

. . . .

. . . .

. . . .

This is a lattice, or arrangement, of numbers.
Imagine that the pattern continues.

1 What number is directly under 27? _____

2 What is the third number from the left in Row 8?

_____

3 What is the last number in Row 10? _____
How do you know?

Name

Lattice Patterns (A)

Goals
- ◆ Identify and continue patterns.
- ◆ Generalize patterns and relationships.

Questions to Ask
- ◆ *What numbers are in Row 5?* (25, 26, 27, 28, 29, 30)
- ◆ *What numbers are in Column D?* (4, 10, 16, 22, 28, . . .)
- ◆ *What do the dots under each column mean?* (The dots indicate that the pattern continues.)
- ◆ *What is the difference between any 2 consecutive column numbers?* (They differ by 6.)
- ◆ *If you continued writing numbers, what would the next number in Column F be?* (36)
- ◆ *What would the last number in Row 7 be?* (42)
- ◆ *How many numbers are in each row?* (6)

Solutions

1 33 is directly under 27.

The numbers increase by 6 in the columns.

2 45 is third from the left in Row 8.

Since Row 7 ends with 42, Row 8 starts with 43. The third number in that row will be 45.

3 60 is the last number in Row 10.

Two possible solution methods:

There are 6 numbers in each row, so there are 60 numbers in the first 10 rows. Therefore Row 10 ends with 60.

The last number in every row is in Column F. All of the numbers in Column F are multiples of 6. Row 1 ends with 1×6, Row 2 ends with 2×6, . . ., Row 10 ends with 10×6.

Algebra: Puzzles and Problems, Grade 6   **105**

Lattice Patterns (B)

Columns

	A	B	C	D	E	F
1	1	2	3	4	5	6
2	7	8	9	10	11	12
3	13	14	15	16	17	18
4	19	20	21	22	23	24
5	25	26	27	28	29	30

Rows

. . . . . .
. . . . . .
. . . . . .

This is a lattice, or arrangement, of numbers.
Imagine that the pattern continues.

1 Which row starts with the number 97?

_____

2 Which row contains the number 50?

_____

How do you know?

3 What column contains the number 75?

_____

How do you know?

Name _____

Lattice Patterns (C)

Columns

	A	B	C	D	E	F
1	1	2	3	4	5	6
2	7	8	9	10	11	12
3	13	14	15	16	17	18
4	19	20	21	22	23	24
5	25	26	27	28	29	30

Rows

. . . . . .
. . . . . .

This is a lattice, or arrangement, of numbers.
Imagine that the pattern continues.

1 What is the ninth number in Column F? _____

2 In which row and column is 125?

_____

3 Which was easier to find, the row or the column that 125 is in? _____
Explain your answer.

Name _____

Lattice Patterns (B)

Solutions

1 Row 17 starts with 97. One way to find the answer is to think: 97 ÷ 6 (the number of numbers in a row) = 16 with 1 left over. That means that Row 16 ends with 16 × 6, or 96, and Row 17 starts with 97.

2 Row 9 has the number 50. One possible solution method is to think: 50 ÷ 6 = 8 with 2 left over. That means that Row 8 ends with 8 × 6, or 48, and 50 will be in Row 9.

3 75 is in Column C. One way to find the answer is to think: 75 ÷ 6 = 12 with 3 left over. That means that 12 × 6, or 72, is the last number in Row 12, 73 is in Column A in the next row, 74 is in Column B, and 75 is in Column C.

Lattice Patterns (C)

Solutions

1 The ninth number in Column F is 54. One way to find the answer is to think: All numbers in Column F are multiples of 6. Row 1 ends with 1 × 6, Row 2 ends with 2 × 6, ..., Row 9 ends with 9 × 6.

2 125 is in Row 21 and Column E. One possible solution method is to think: 125 ÷ 6 = 20 with 5 left over. That means that Row 20 ends with 20 × 6, or 120. Row 21 starts with 121 in Column A, so 125 is the fifth number in that row, in Column E.

3 It is probably easier to find the column number than the row number. Once you know which row 125 is in, it is easy to determine the column.

Lattice Patterns (D)

Columns

Row 1	2				
Row 2	2	4			
Row 3	2	4	6		
Row 4	2	4	6	8	
Row 5	2	4	6	8	10

.   .   .   .   .   .

.   .   .   .   .   .

.   .   .   .   .   .

This is a lattice, or arrangement, of numbers.
Imagine that the pattern continues.

1 What is the last number in Row 10? _____

2 In which row is 60 the last number? _____

3 Explain how to find the last number in a row.

Name

108   **Algebra: Puzzles and Problems**

© Creative Publications

Permission is given by the publisher to the purchasing teacher or parent to reproduce this page for classroom or home use only.

Lattice Patterns (D)

Goals
- ◆ Identify and continue patterns.
- ◆ Generalize patterns and relationships.

Questions to Ask
- ◆ *What numbers are in Row 5?* (2, 4, 6, 8, 10)
- ◆ *What is the last number in Row 4?* (8)
- ◆ *What do the dots under each column mean?* (The pattern continues. There are more rows of numbers.)
- ◆ *What numbers will be in Row 6?* (2, 4, 6, 8, 10, 12)

Solutions

1 The last number in Row 10 is 20.

2 60 is the last number in Row 30.

3 By looking at row numbers and the last numbers, you can figure out the last number in any row. The last number in a row equals two times the Row Number.

(Last Number = 2 × Row Number)

Row Number	1	2	3	4	R
Last Number	2	4	6	8	$L = 2R$

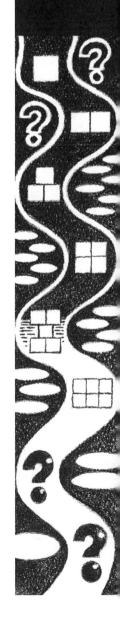

Lattice Patterns (E)

Columns

Rows										
1	2	4								
2	2	4	6	8						
3	2	4	6	8	10	12				
4	2	4	6	8	10	12	14	16		
5	2	4	6	8	10	12	14	16	18	20

. . . .
. . . .
. . . .

This is a lattice, or arrangement, of numbers.
Imagine that the pattern continues.

1 How many numbers are in Row 40? _____
How do you know?

2 What is the last number in Row 30? _____

3 Write a rule for finding the last number in any row.

Lattice Patterns (F)

Columns

Rows									
1	2								
2	2	4	6						
3	2	4	6	8	10				
4	2	4	6	8	10	12	14		
5	2	4	6	8	10	12	14	16	18

. . . .
. . . .
. . . .

This is a lattice, or arrangement, of numbers.
Imagine that the pattern continues.

1 How many numbers are in Row 20? _____

2 What is the last number in Row 20? _____

3 Write a rule for finding the last number in any row.

Lattice Patterns (E)

Solutions

1 There are 80 numbers in Row 40. Make a table showing the row numbers and the number of numbers in a row, to find the answer.

Row Number	1	2	3	4	R
Number of Numbers	2	4	6	8	2R

Number of Numbers in a row = 2 × Row Number.

2 The last number in Row 30 is 120.

3 The table shows the Row Number and Last Number in each row of the lattice.

Row Number	1	2	3	4	R
Last Number	4	8	12	16	4R

Last Number = 4 × Row Number.

Lattice Patterns (F)

Solutions

1 There are 39 numbers in Row 20.

2 The last number in Row 20 is 78.

3 Make a table to show the relationship between the last number and the row number.

Row Number	1	2	3	4	R
Last Number	2	6	10	14	4R − 2

Differences between consecutive numbers in the Last Number column are always 4. That means that part of the rule is, Multiply by 4. Multiply row numbers by 4 and compare the products with the last numbers. In all cases, you must subtract 2 from the product to get the last number. 4 × Row Number − 2 = Last Number.

Algebra: Puzzles and Problems, Grade 6

Shape Teaser (A)

Shape Number	1	2	3	4
				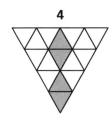

This is a pattern of growing shapes.
Imagine that the pattern continues.

Shape Number	1	2	3	4
Black Triangles	1	2	3	4
White Triangles	0	2	6	12

How many black triangles will there be in:

1 Shape 10? _____

2 Shape 30? _____

3 Write a rule for finding the number of black triangles in a shape. Use the shape number in your rule.

How many black triangles and white triangles altogether will there be in:

4 Shape 10? _____

5 Shape 30? _____

6 Write a rule for finding the number of black triangles and white triangles in a shape. Use the shape number in your rule.

Name

Shape Teaser (A)

Goals
- ◆ Identify and continue patterns.
- ◆ Write a rule that relates a shape to its location in a sequence.

Questions to Ask
- ◆ *How many black triangles are in Shape 3?* (3)
- ◆ *What is the total number of small triangles, black and white, in Shape 3?* (9)
- ◆ *Describe Shape 5. How many black triangles do you think it will have?* (5)
- ◆ *How many small triangles, black and white, are in Shape 4?* (16)
- ◆ *How many small triangles do you think there will be in Shape 5?* (25)
- ◆ *What can you say about 1, 4, 9, 16, the numbers of small triangles?* (They are square numbers.)

Solutions

1 10 black triangles

2 30 black triangles

3 The number of black triangles in a shape is the same as the Shape Number. Let S represent the Shape Number and B represent the number of black triangles. Then $B = S$.

4 100 triangles

5 900 triangles

6 The number of small triangles (black and white) in a shape is equal to the Shape Number times itself, or the square of the Shape Number. Let T represent the number of small triangles. Then $T = S \times S$, or $T = S^2$.

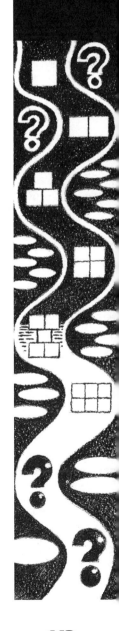

Algebra: Puzzles and Problems, Grade 6

Shape Teaser (B)

Shape Number   **1**   **2**   **3**   **4**

0 black   1 black   2 black   3 black
1 white   3 white   5 white   7 white

This is a pattern of growing shapes. Imagine that the pattern continues. How many white triangles will there be in:

1 Shape 8? _____

2 Shape 100? _____

3 Write a rule for finding the number of white triangles in a shape. Use the shape number in your rule.

How many black triangles will there be in:

4 Shape 8? _____

5 Shape 100? _____

6 Write a rule for finding the number of black triangles in a shape. Use the shape number in your rule.

Permission is given by the publisher to the purchasing teacher or parent to reproduce this page for classroom or home use only.

Shape Teaser (C)

   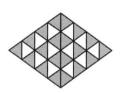

Shape Number   **1**   **2**   **3**   **4**

2 black   6 black   12 black   20 black
0 white   2 white   6 white   12 white

This is a pattern of growing shapes. Imagine that the pattern continues. How many black triangles will there be in:

1 Shape 8? _____

2 Shape 10? _____

3 Write a rule for finding the number of black triangles in a shape. Use the shape number in your rule.

How many white triangles will there be in:

4 Shape 8? _____

5 Shape 10? _____

6 Write a rule for finding the number of white triangles in a shape. Use the shape number in your rule.

Permission is given by the publisher to the purchasing teacher or parent to reproduce this page for classroom or home use only.

Shape Teaser (B)

Solutions

1 15 white triangles

2 199 white triangles

3 The total number of white triangles is equal to the Shape Number plus the Shape Number minus one, or two times the Shape Number minus one.

Let W represent the number of white triangles. Then $W = S + S - 1$, or $W = 2S - 1$. (For example, in Shape 3, there are S white triangles in the top row and $S - 1$ white triangles in the bottom row, a total of $S + S - 1$ white triangles.)

4 7 black triangles

5 99 black triangles

6 The number of black triangles is equal to the Shape Number minus one. $B = S - 1$.

Inductive Reasoning

Shape Teaser (C)

Solutions

1 72 black triangles

2 110 black triangles

3 The number of black triangles is equal to the Shape Number times the number that is one more than the Shape Number. $B = S \times (S + 1)$. Or, multiply the Shape Number by itself and add the Shape Number, $B = S \times S + S$.

4 56 white triangles

5 90 white triangles

6 The number of white triangles is equal to the Shape Number times the number that is one less than the Shape Number, $S \times (S - 1)$, or the Shape Number squared minus the Shape Number, $S^2 - S$.

Inductive Reasoning

Shape Teaser (D)

This is a pattern of growing rectangles.
Imagine that the pattern continues.

Rectangle Number	1	2	3
Small Squares	2	6	12
Black Squares	1	3	6
White Squares	1	3	6

How many small squares will there be in:

1 Rectangle 5? _____

2 Rectangle 10? _____

3 Write a rule for finding the number of small squares in a
rectangle. Use the rectangle number in your rule.

How many black squares will there be in:

4 Rectangle 5? _____

5 Rectangle 10? _____

6 Write a rule for finding the number of black squares in a
rectangle. Use the rectangle number in your rule.

Name

Shape Teaser (D)

Goals
- ◆ Identify and continue patterns.
- ◆ Write a rule that relates a shape to its location in a sequence.

Questions to Ask
- ◆ *How many small squares are in Rectangle 3?* (12)
- ◆ *How many black squares are in Rectangle 3?* (6)
- ◆ *How many small squares will there be in Rectangle 4? (20) How do you know?* (Rectangle 4 will be 4×5.)
- ◆ *How many black squares will there be in Rectangle 4?* (10) *Explain your answer.* (The number of black squares seems to be half the number of total squares.)

Solutions

1 30 small squares

2 110 small squares

3 The number of small squares in a rectangle equals the Rectangle Number times one more than the Rectangle Number.
[Rectangle Number $\times$ (Rectangle Number + 1)] or
(Rectangle Number^2 + Rectangle Number)

4 15 black squares

5 55 black squares

6 The number of black squares in a rectangle equals half the product of the Rectangle Number and one more than the Rectangle Number.
B = (one half) [(Rectangle Number) $\times$ (Rectangle Number + 1)]

Shape Teaser (E)

Rectangle Number

1	2	3
6 Total	12 Total	20 Total
0 Black	2 Black	6 Black
6 White	10 White	14 White

This is a pattern of growing rectangles. Imagine that the pattern continues.

How many black squares will there be in:

1 Rectangle 5? _____

2 Rectangle 10? _____

3 Write a rule for finding the number of black squares in a rectangle. Use the rectangle number in your rule.

How many white squares will there be in:

4 Rectangle 5? _____

5 Rectangle 10? _____

6 Write a rule for finding the number of white squares in a rectangle. Use the rectangle number in your rule.

Shape Teaser (F)

Rectangle Number

1	2	3
6 Total	12 Total	20 Total
4 Black	6 Black	8 Black
2 White	6 White	12 White

This is a pattern of growing rectangles.

Imagine that the pattern continues.

How many black squares will there be in:

1 Rectangle 5? _____

2 Rectangle 10? _____

3 Write a rule for finding the number of black squares in a rectangle. Use the rectangle number in your rule.

How many white squares will there be in:

4 Rectangle 5? _____

5 Rectangle 10? _____

6 Write a rule for finding the number of white squares in a rectangle. Use the rectangle number in your rule.

Algebra: Puzzles and Problems

Shape Teaser (E)

Solutions

1 20 black squares

2 90 black squares

3 The number of black squares in a rectangle is equal to the Rectangle Number times one less than the Rectangle Number.

$B = (\text{Rectangle Number}) \times (\text{Rectangle Number} - 1)$

4 22 white squares

5 42 white squares

6 The number of white squares in a rectangle is equal to four times the Rectangle Number plus two.

$W = (4 \times \text{Rectangle Number} + 2)$

Shape Teaser (F)

Solutions

1 12 black squares

2 22 black squares

3 The number of black squares in a rectangle is equal to two times one more than the Rectangle Number.

$B = [2(\text{Rectangle Number} + 1)]$

4 30 white squares

5 110 white squares

6 The number of white squares in a rectangle is equal to the Rectangle Number times one more than the Rectangle Number.

$W = [(\text{Rectangle Number}) \times (\text{Rectangle Number} + 1)]$

Certificate of Excellence

in Algebra

This is to certify that

_____

has satisfactorily completed all the problems for the big idea

_____

and is considered to be an expert in Algebra.

Date _____

School _____

Grade _____

Teacher _____